Developing and Sustaining Excellent Packaging Labelling and Artwork Capabilities

Delivering patient safety, increased return and enhancing reputation

Stephen McIndoe

Andrew Love

Developing and Sustaining Excellent Packaging Labelling and Artwork Capabilities
Delivering patient safety, increased return and enhancing reputation

First published in 2012 by
Ecademy Press
48 St Vincent Drive, St Albans, Herts, AL1 5SJ
info@ecademy-press.com
www.ecademy-press.com

Printed and bound by Lightning Source in the UK and USA

Designed by Julie Oakley

Printed on acid-free paper from managed forests. This book is printed on demand, so no copies will be remaindered or pulped.

ISBN 978-1-908746-16-0

The right of Stephen McIndoe and Andrew Love to be identified as the authors of this work has been inserted in accordance with sections 77 and 78 of the Copyright Designs and Patents Act 1988.

A CIP catalogue record for this book is available from the British Library. All rights reserved. No part of this book may be reproduced in any material form (including photocopying or storing in any medium by electronic means and whether or not transiently or incidentally to some other use of this publication) without the written permission of the copyright holder except in accordance with the provisions of the Copyright, Design and Patents Act 1988. Applications for the Copyright holders written permission to reproduce any part of this publication should be addressed to the publishers.

This book is available online and in all good bookstores.

© 2012 Stephen McIndoe and Andrew Love

This publication is for information purposes only, comprises the opinions of the authors, and does not constitute professional advice on the matters discussed. Any action you take in reliance on it is entirely at your own risk. You should first seek advice from a qualified professional familiar with your individual circumstances. The information in this publication is not intended to be comprehensive, and many details which may be relevant to particular circumstances have been omitted. Accordingly it should not be regarded as a complete and authoritative source of information on the subject matter. We expressly disclaim any and all liability or responsibility for the consequences of any inaccuracies or omissions. Please visit our website at www.be4ward.com or contact us at enquiries@be4ward.com to find out how we can help you in your specific business circumstances.

Dedication

To all those people dedicated to improving patient safety and to our families who we love dearly for supporting us though this great journey.

Acknowledgements

To the many colleagues and clients who we have worked with in the past, without whom none of this could have been possible.

Contents

		page
	Foreword	1
1	Why artwork capabilities are key to business success	5
2	How artwork errors happen	17
3	Service orientation	25
4	The core artwork process	35
5	Interfacing processes	69
6	Supporting processes	73
7	Organisation	95
8	Technology	109
9	Outsourcing	145
10	Future developments	161
11	Making it happen	175
	Close	195
	Glossary of Terms	199
	About the Authors	203

Foreword

Packaging artwork is an often forgotten back room process in most pharmaceutical companies, but the changing business environment has brought issues from this capability to the fore.

Pharmaceutical and other healthcare companies are facing one of the most difficult periods in their history. Current products are rapidly going off patent leaving significant revenue challenges. At the same time, weak product pipelines are failing to fill the gap.

Furthermore, global markets are changing rapidly. Traditional markets are stagnating and new markets are evolving at a rapid pace. Everywhere, key healthcare purchasers are putting increasing pressure on drug prices.

In response to these significant challenges, pharmaceutical companies are looking to make the most out of their current assets. This often manifests itself in a drive to launch as many product variants in as many markets as possible. For the traditional molecule-based global pharmaceutical companies, this represents a significant change in strategy.

The rapid growth in the number of drugs coming off patent, together with the increasing pressure on price from the major purchasers, has led to a huge opportunity and growth for generic pharmaceutical companies. For them the challenges are very similar to the pharmaceutical companies, namely to market as many product variants in as many markets, as quickly as possible.

In today's world, all drug companies have an increasing need to develop and maintain an excellent reputation with a diverse group of stakeholders. Pharmaceutical companies are looking to develop and maintain trust with governments and purchasing groups in order to help maintain the product prices necessary to support their significant drug development spending. The increasing competition amongst generic companies means that they each need to develop and sustain their reputation in order to win business and maintain their production licences.

Maintaining this reputation whilst rapidly growing the number of products is particularly challenging when one considers that the largest single cause of product recall is due to packaging errors. Recognising this, regulators around the world are focusing on driving improvement in all business capabilities associated with the management of packaging design and manufacture.

When launching product variants in new markets, much if not all of the physical packaging design is already established. The text and graphics, or artwork as it is known, that is placed on these physical components is what changes every time. It is this artwork design and maintenance capability that becomes critical to achieving and maintaining the objectives of both pharmaceutical and generic drug companies.

For a large global pharmaceutical company, developing artwork for tens of thousands of products is typically a process involving thousands of people, in over a hundred countries, from tens of different organisations. To orchestrate all of this activity, the right combination of business processes, organisation design, information technology, facilities and suppliers has to be managed.

For smaller organisations, whilst the scale of the problem may be reduced, all of the same challenges have to be met.

This book describes the key capabilities required to deliver right-first-time packaging artwork in today's environment. It also discusses potential future developments in the area to help the reader design any improvement activity with these in mind.

Finally, it looks at how an organisation can go about understanding how they need to adapt and improve their capabilities to meet their evolving business strategy and go about the often complex change-management journey to achieve it.

Steve Richmond, *Head of Global Packaging, AstraZeneca*

1
Why artwork capabilities are key to business success

Effective packaging labelling and artwork development capabilities are critical to the success of pharma companies in today's new world.

Today, large pharmaceutical companies are arguably in the midst of the toughest challenges that have ever faced them. Blockbuster drug patents are expiring at an alarming rate and product development pipelines are not ready to fill the ensuing gap; governments and other powerful purchasing groups are demanding ever lower prices for pharmaceutical products; regulators and the public are increasingly sceptical about the role of pharmaceutical companies in looking after the public's welfare; regulatory authorities are becoming more and more stringent in their definition and enforcement of regulations; competition from generic companies is becoming fierce; they are saddled with a very high infrastructure cost born out of the time when supply-chain cost did not matter and getting the new blockbuster products to market securely was the main business priority.

To respond to these formidable challenges, pharmaceutical companies appear to be focusing their activities in a number of areas.

For the medium to long term, companies need to strengthen their product pipelines. They are achieving this through a combination of internal research capability development and, to an increasing extent, encouraging broad external research. The key to this latter strategy is to develop the capability to identify promising new drugs at the optimal point in their early development lifecycle and purchase the rights to them at a reasonable cost. All this activity requires a significant amount of high-risk funding.

Therefore, in the short term, pharmaceutical companies are focusing their attention on maximising the value that they can extract from their current assets, whilst at the same time restructuring their high cost-base. One of the significant ways they can achieve this is to create and launch as many existing product variants, in as many markets, through as many channels as possible. For many companies this represents a significant change of strategy as traditionally they would have focussed their attention on a few large products in the larger volume markets.

At the same time as pharmaceutical companies try to maximise sales volume, they are also working hard to maintain as high a selling price as possible, particularly in the wealthy western and emerging markets. One significant strand of activity which supports this goal leads them to focus on developing trust with key stakeholders that their intentions are for mutual good, and not just for their own profit. This in turn drives them to a greater focus on meeting regulatory requirements and being seen publicly to do the right thing. They need to be increasingly vigilant in this area as the rise of the global information superhighway means that stakeholders anywhere in the world get to know bad news immediately.

At the same time that the large pharmaceutical companies struggle to find their place in the new order, many opportunities are generated for smaller players, particularly in the areas of biotechnology and vaccines. Here, the focus is on getting relatively short development cycle products to as many markets as quickly as possible, in a manner that does not compromise patient safety.

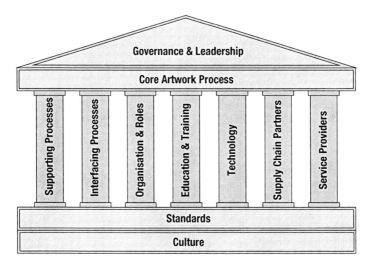

Figure 1.1 Typical labelling and artwork management capability elements

At the heart of much of the activity described above is the ability to develop, launch and maintain new products and variants in many different markets as rapidly as possible, whilst at the same time ensuring that all regulatory requirements are demonstrably met. Achieving this can only be done if a company puts in place an effective packaging labelling and artwork management ability. This book will help you understand how to achieve this and therefore drive the success of your business.

Unfortunately, for too many pharmaceutical companies, the reality of their current packaging labelling and artwork development capability falls short of what is required to meet these new business challenges. This is most manifestly obvious when one considers that errors in packaging and labelling are reported as the single largest source of product recalls today.

The manufacture and supply of pharmaceutical products is a complex business involving many participants, specialist skills and technologies. It is therefore perhaps surprising that packaging and labelling errors present such an issue. Furthermore, this poor performance appears to have the increasing attention and scrutiny of regulatory agencies worldwide.

As many who have been involved in managing or trying to rectify issues in this artwork space will testify, it presents its own set of unique complexities that are far from trivial to solve.

Defining some terms

To simplify the terms used in this book, from this point on we will use the term "artwork" to mean all aspects of artwork and labelling. The term describes any text or graphics which are applied to packaging components such as labels, cartons and patient information leaflets.

We also make extensive use of the term "capability". By this we mean all those elements which need to come together to ensure that an end result is delivered in a sustainable way. In the case of artwork capabilities, this will include such aspects as business processes, people, IT tools and third-party service providers.

For definitions of other terms used in this book, please refer to the glossary of terms.

Why should you be reading this book?

You may be someone who is looking to ensure that your artwork capability can help support your company's new business strategy in the most effective way; you may be a senior manager responsible for this capability and need to understand the topic more thoroughly; you may be someone who is currently dealing with the rectification of a packaging labelling and artwork error and looking for guidance on how to make improvements; you may be looking to avoid errors and make your capabilities more robust; you may just be curious about the topic and wish to learn more.

Whatever the reason, this book aims to explain the elements which make up a comprehensive artwork management capability within a typical pharmaceutical company and explains how you can go about making improvements to your existing capability in a managed way.

What is an artwork error?

We discuss the types of artwork errors and their causes in more detail in the next chapter. There we categorise artwork errors into four groups:

Gross errors
Where significant information is omitted. An example would be completely missing the need to change a piece of artwork in response to a new regulatory requirement.

Context and meaning errors
Where information is presented in an ambiguous or incorrect way on the artwork. An example of this might be the inappropriate use of hyphenation causing ambiguous or incorrect meaning.

Content errors
Where there are errors and omissions in the detailed content of the artwork. An example of this would be incorrect symbols being used in the artwork.

Technical errors

Where there are errors or omissions in the technical aspects of the artwork. An example of this would be the wrong specification of barcode being used in an artwork.

The implications of an artwork error

The implications of an artwork error can be as far-reaching and serious as any other error with the supplied product. Artwork text and graphics describe the product and provide information and instruction for its safe and effective use.

Impact on patients

The bond between the patient and their medicine is deep-rooted. Patients trust that the product will make them better and expect that it has been developed, manufactured and supplied to the highest quality and ethical standards. Errors in the information provided with the product are significant and can be life-threatening. We are sure that you will agree that this risk to the patient's well-being is not acceptable and their confidence in the treatments they are taking must be maintained. Trust is easily lost and almost impossible to recover.

Impact on prescribers

All prescribers (whether doctors, pharmacists, nurses or other healthcare professionals) are busy people with a clear mission – to make the patients they treat better. They expect that the products and information they are provided with are fit for purpose, error-free and safe to use. They don't want to administer products that will make their patients sicker. Rectifying the patient issues created by artwork errors is a burden they neither want nor welcome. Furthermore, the remedial action following an incident diverts their limited resources away from their core purpose.

These healthcare professionals are often the final decision-makers when it comes to selecting the product that is prescribed or used in the future. Hence, any lack of confidence that they may have in a particular product, brand or company can have a direct impact on the products that get used.

Also, it must not be forgotten that there is also a serious personal impact for some prescribers involved in incidents leading to patient harm. Indeed, a number of prescribers involved in such incidents go on to leave their chosen profession altogether.

Impact on regulators

The remit of the pharmaceutical regulators, amongst other things, is to set and enforce the standards by which the industry must operate to ensure patient safety. They have the authority to allow or block product use and the power to take punitive action against companies who they see fail to meet expected standards. The regulatory environment is becoming ever more complex and stringent and there is less and less tolerance for artwork error. Moreover, as we have already observed, the information age means that an incident in any country has visibility to all regulators worldwide.

It is therefore understandable that regulators expect companies to be continually striving to eliminate artwork errors, and take appropriate actions to reinforce that view.

Impact on pharmaceutical company staff

Two groups of pharmaceutical company staff are typically impacted by an artwork error: the team managing the recall and the operations teams who support the artwork process in which the error occurred.

The team managing the recall need to focus on the immediate and urgent tasks related to identifying the impacted product, withdrawing it from the supply-chain and reinstating adequate supply as quickly as possible. Whilst challenging, this work is more often than not very motivating for those involved as a great deal of satisfaction can be derived from solving the immediate and significant recall problem.

The impact on the staff involved in the operation of the artwork process is somewhat different. Not only are they likely to be involved in the rectification activity, they will be heavily involved in the incident enquiry and corrective and preventative actions. Furthermore, there are the undoubted performance and morale issues that will likely need to be addressed.

Impact on the company

The impact on the company can be significant. The patient safety implications are counter to any pharmaceutical company's core values. This is compounded by the sales, reputation and sanction impacts, through unfavourable publicity, loss of customer confidence, possible loss of licence and increased regulator scrutiny and action. As we discussed earlier, in today's business environment, these impacts are potentially significant to the success of the company.

The cost impacts of these errors are also substantial. There are the immediate tangible costs of recall, product write-off, repacking and market re-supply. However, these can be overshadowed by the less tangible follow-on costs occurring through loss of sales and market share, customer reimbursement and litigation. In the extreme these not only impact the bottom line, but can directly influence the company's share price.

How are errors discovered?

Artwork errors will inevitably occur in the process and the hope is that when they occur, they are picked up internally by the various quality control checks in the process. Typically, internal checks would occur at several points in a normal artwork process.

However, when these processes fail, the source of detection will be external. This is usually by people who interact with the product packaging – typically prescribers, healthcare professionals administering the product or patients using the product. Errors can be detected in many ways, depending on their nature:

- By chance, seeing a discrepancy that no-one else had noticed.
- Through the diligence of medical professionals checking the product before dispensing or use.
- By identifying an issue with the labelling or instruction when they went to use the product.
- By using the wrong product (e.g. strength or dosage) or using the product in the wrong way and witnessing an adverse medical reaction.

As discussed before, all of these errors have potentially serious consequences for the safety and health of the patient and trust and acceptance of the product.

Actions that need to be taken when an error is discovered

An externally identified error will normally result in a combination of the following:

- A customer complaint to the pharmaceutical company.
- A report of an adverse medical reaction to the pharmaceutical company.
- A complaint to the local pharmaceutical regulator.

Whilst what happens next is often a series of parallel activities, we have described them sequentially for simplicity.

The first step that needs to be undertaken is an assessment of the severity of the risk and whether a product recall is required to protect patient safety. The magnitude of the recall also needs to be considered – distribution level, pharmacy level or patient level – and again this will be determined based upon the risk to patient safety.

Relevant external regulators need to be informed, a recall team established and the recall commenced as per the company's recall procedure. The first stage will be to undertake necessary corrective actions, tracing the relevant batches and communicating to wholesalers, distributors, pharmacists and doctors as necessary. In addition, all internal finished, work-in-progress and packaging material stocks need to be quarantined.

Once the supply-chain is secure and relevant product returned as appropriate, action must be taken to provide replacement stock to the marketplace.

This may require new stocks to be packed, existing product to be re-packed or existing product to be over-labelled. This will depend on the type of error and the type of packaging material involved. This is also likely to require new artwork and components to be rapidly expedited – either for the new packaging materials or for over-labels. The company will need to decide whether this rectification is being undertaken near the marketplace or at the original packaging site (if these are different).

Once the market has been re-supplied, the task of understanding what has gone wrong and undertaking preventative actions can commence.

To ensure that the cause of the error is correctly understood and the correct actions are undertaken it is essential that the investigation concentrates on identifying the root causes of the incident. In the environment likely at this time, it is too easy to blame individuals rather than see the error as an opportunity to learn and improve the process as a whole. If the root causes are really understood, they can be designed out of the process so they cannot happen again. This could be through error-proofing, improved instruction, checklists, improved process steps or enhanced training. The learning needs to be shared across all relevant parts of the organisation to ensure that capabilities are appropriately improved.

This is a golden opportunity for the rigorous application of continuous improvement tools and behaviours.

Identified preventative actions need to be assigned to appropriate resolving individuals and teams and tracked to ensure they are addressed in a timely way. This would typically be through the company's corrective action and preventative action tracking.

Depending on the seriousness of the incident, there may be further action required by the external regulator. This may include location or process audits, supply restrictions, product licence withdrawals or other sanctions

All of the above places additional strain on the organisation. Many teams are impacted by the activity which, by nature, is unplanned. There is an additional leadership action therefore required to monitor the resilience of teams and individuals, set the appropriate climate for the investigation and tone for communications and ensure the effectiveness of day-to-day operational activity is not adversely impacted (leading to further quality impacts). When returning to routine operation, leadership will then need to support rebuilding team morale and raising the organisation's capability as necessary.

It can be correct

Many pharmaceutical companies have been tackling the problem of errors in their artwork processes. Through careful process design and the establishment of necessary underpinning capabilities and competency, performance can be dramatically improved.

In the following chapters in this book we will discuss the process and capability improvements in more detail, outlining improvements to be made and plans to address.

As already discussed, achieving excellence in this area can help deliver many significant strategic benefits:

- Increased patient safety.
- Improved regulatory compliance.
- Increased sales.
- Improved profit margin.
- Improved reputation.
- Reduced cost and valuable resource absorption.

In the rest of this book we will discuss the capabilities which, in the authors' experience, are necessary to establish a sustainable artwork capability that can deliver these benefits for an organisation. We will also discuss the approaches that companies can take to improving their existing artwork activities.

2
How artwork errors happen

Before we start to examine how to address the issues discussed in the first chapter, it is valuable to look at where information on packaging comes from, the types of errors that typically occur and discuss some of the ways they occur. Once we have this understanding, we will be better placed to discuss the capabilities required to deliver right-first-time artwork.

The sources of artwork content

Product packaging artwork is the manifestation of requirements from many different stakeholders, both within a pharmaceutical company and externally to it. The resulting artwork is often a complex mix of multilingual text and graphical elements. In the next few pages, we will examine the primary sources of this information.

External health authorities are clearly key to shaping the requirements for the content of artwork. Their primary role is to work with a pharmaceutical company to agree the medical text that needs to be placed on the different components of the artwork. If we consider a patient information leaflet as an example, the vast majority of this component derives from the medical text that was agreed with the external health authority. In contrast, as can be seen in Figure 2.1, a carton contains information from a broader range of sources.

Other external legislators will also play a part in shaping the requirements for packaging artwork. As an example, many countries are increasingly introducing legislation to drive recycling as one of their solutions to help protect the environment. In many cases, pharmaceutical products are not exempt from these pieces of legislation and therefore pharmaceutical product packaging must comply. This often means that packaging artwork needs to carry country and/or internationally recognised recycling symbols or text.

Packaging clearly contains information to help patients and medical professionals administer the products it contains. By careful design of the information and packaging, a pharmaceutical company has an opportunity to improve adherence and persistence.

Product packaging, along with many other information sources, is used by the sales and marketing functions of a pharmaceutical company to emphasise and build the product brand and drive sales. It is important that elements such as logos, trade names and graphical brand images are portrayed on the packaging in a consistent way. However, it is not uncommon for these requirements to run at odds with the requirements of the external health authorities. Sometimes, the packaging design process is the place where these differing design drivers manifest themselves for the first time.

It goes without saying that packaging design is driven by the need to deliver the product to the patient in a safe and effective way. Therefore, the requirements defined during product development need to be incorporated in any packaging design. It is worth noting that it is important to understand these requirements and their potential impacts early in the product packaging development cycle as, once defined and submitted in the product registration and approval package, they can be very difficult to change.

Supply-chain requirements – both within the pharmaceutical manufacturing company and downstream – should be taken into account in the packaging design. At one level this may be the inclusion of

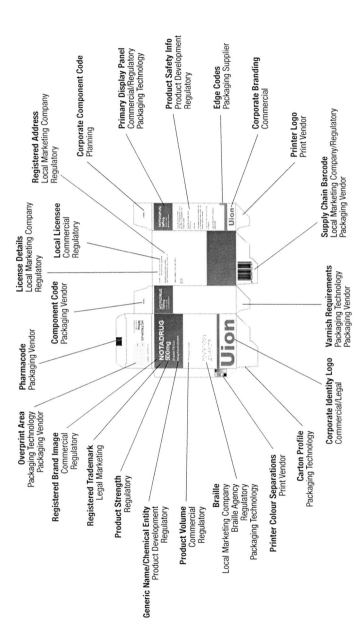

Figure 2.1 Typical sources of carton information

machine-readable coding for use in automated packaging, distribution and dispensing operations. At another, the fundamental pack design may need to change to take account of requirements such as postal product distribution.

Other groups, such as the legal department, are also likely to need to provide input to the packaging design through the supply of, for example, trademark information.

Figure 2.1 shows an example of a typical carton and illustrates the sources of the information which make up the artwork.

Artwork errors

For the purposes of the discussion here, we will divide artwork errors into three broad categories:

Gross errors

This type of error occurs when mistakes are made impacting the whole or large portions of the artwork. At the highest level this will include omitting to change a complete artwork when required. Other errors in this category would include missing complete paragraphs of information, or incorrectly substituting the wrong language translation into parts of the artwork text.

Context and meaning errors

In this type of error, the information provided on the artwork is in some way misleading to the user, such as: inappropriate use of punctuation; layout causing meaning to be confusing; mistranslation. This can happen on individual panels of an artwork, as well as being caused by the interaction of different panels of an artwork as they form the 3D packaging component object.

Content errors

This is the general category of errors which captures issues such as: missing or incorrect words or characters, typos and misspellings; missing some information which should be repeated in a different part of the artwork; incorrect positioning of information; using incorrect graphics such as brand images and logos.

Technical errors

Technical errors cover issues such as: the use of the wrong component layout; size errors; errors in colour identification; issues with barcode and Braille specifications; incorrect varnish specification.

Causes of artwork errors

We have divided the many causes of artwork errors into a number of categories.

Process gaps and inconsistencies

Alternatively termed as systematic errors, these occur when the design of the business process is incomplete or conflicting, leading to errors in the content of the artwork. A typical example of this would be a gap in the process definition for the provision of a particular piece of information.

Lack of competence

Here, operators do not have the necessary skills, knowledge or instructions to carry out the tasks that are required of them in the business process. This may be due to issues such as an inadequate level of process definition or inadequate training and competence assessment.

An issue of particular concern in artwork processes which we discuss later is that of ensuring the competence of people who only perform tasks in the process very infrequently.

Inappropriate decision-making

In this type of situation, people will make inappropriate decisions during the execution of the business process which lead to errors in the resulting artwork. For example, management may set priorities which are interpreted by operations staff to put moving an artwork to the next stage of the process ahead of doing a task completely and correctly.

Ambiguity

The artwork process involves many individuals providing detailed instructions to other individuals in the process, with the resulting opportunity for ambiguity in these instructions to lead to errors in the artwork. A lack of templates or instructions on how to pass on information and instructions in an unambiguous way can be examples of this type of issue.

It must be remembered that many people working in the artwork process do so in their second language. This in itself significantly increases the possibility of individuals misinterpreting instructions which are not entirely clear.

Errors in source information

The age-old phrase "garbage in, garbage out" applies very well to the artwork process. If incorrect source information is used in the process then it is highly likely to cause errors in the resulting artwork. Typical examples of this type of issue include: people using the wrong or incorrect versions of documents; the use of uncontrolled information sources such as ad hoc personal spreadsheets.

Human error

A typical artwork process includes many steps where people are directly responsible for carrying out activities such as: transcribing information from one source to another; performing multiple complex or repetitive tasks. It is the nature of human beings that we make mistakes for many reasons. Sometimes it will be due to limitations described elsewhere in this section, sometimes it may just be because we are having a bad day. Whilst many steps can be taken to help reduce the possibility of human error, the fact remains that it can still happen and needs to be taken account of when designing artwork capabilities.

Technology errors

Technology in the form of computer software and tools is often used to perform or aid many artwork process steps. However, without careful design and control, this technology can introduce errors into an artwork. Examples of the types of issues which may cause such errors include: software operating incorrectly; systems not providing the user with a true image of a particular document; font transcription errors when moving information from one document to another.

Now that we have looked at the types of errors which occur and their causes, we are in a position to start to focus on how to prevent them and create right-first-time packaging artwork.

3

Service orientation

The development of packaging labelling and artwork involves many different groups across the company and, more often than not, external service providers and supply-chain partners. Figure 3.1 outlines these impacted stakeholder groups.

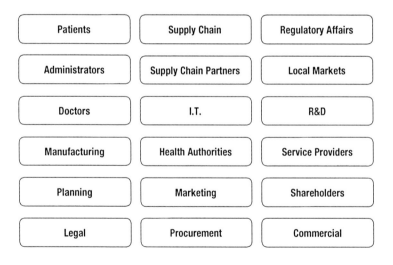

Figure 3.1 Typically impacted stakeholder groups

As we have already discussed, the creation of artwork requires many elements of information to be drawn together in a way that ensures that every detail is correct in the end result.

Without careful orchestration, the separate groups – both within and external to the company – involved in the artwork creation process will not deliver artwork of the required quality standard. Each person involved in the process must perform their task in the process in the correct sequence, using the right information and tools in order to achieve a quality result.

To facilitate this, it is beneficial to consider the provision of labelling text and artwork as a business service. In our experience, the best artwork capabilities are those that consider them to be providing a service to the key business stakeholders and strive to understand their service role and deliver it. Like any service offering, this will evolve over time as the customer's needs change. The management of the artwork capability should recognise these changes and adapt the service accordingly in a managed and considered way.

The development of a clear mission, vision and performance measures can go a long way to orchestrate the successful delivery of the service across the diverse groups that are involved. Figure 3.2 outlines typical elements of such a mission and vision for an artwork service.

Figure 3.2 Typical mission and vision elements

Defining service requirements

When designing an artwork service, we have found it useful to take a systematic approach to the definition of the service requirement based on a number of key questions which we will now discuss.

What is the service producing?

In this case you may conclude that the service is producing accurate packaging components, with two intermediate products, labelling text and packaging artwork.

What is the scope of the service?

Does it include all products in all markets? Does it only cover certain divisions or territories within the company? Do you include external service providers (e.g. translation providers, artwork houses, print suppliers, contract packaging operations)? Does it cover all printed packaging components? In terms of end-to-end process, where does the service start and finish? Does it include other types of product labelling and packaging (e.g. clinical trials' artwork and packaging, promotional information, web-based product labelling)? Does it include the definition of information printed in the packaging operations? Does it include the management of codes for products and components?

It is important to be clear about what is included in the service and what is excluded so that everyone involved is clear on the boundaries they are working within and what to do for situations outside those boundaries.

Who are the customers?

This is a deceptively complex question.

There are clearly external customers who are involved in the use of the products:

- Patients – who clearly expect a product which is effective and safe. Increasingly, they also demand products which are easier to use and help them ensure compliance and persistence.

- Doctors and other healthcare providers who administer the product – their requirements often focus on needing products which are designed for easy identification and administration. This is particularly challenging in the context of large healthcare facilities like hospitals.

- Dispensaries – here the requirements tend to focus on having products which are designed to ensure that the right product is picked for a particular prescription and that it is efficient to handle once selected.

- Prescribers – in the context of artwork design, prescribers are primarily interested in having accurate, clear and concise prescribing information available to them.

Other external customers include:

- Medicines regulators who approve the products.
- Other regulatory bodies who define such things as packaging requirements.
- Purchasers who pay for the products.
- Customer supply-chains which have to deliver the product to the ultimate users.

A small selection of internal customers includes the following groups, and there are many more:

- The person who requests a change to the packaging. This could be commercial, regulatory, engineering or manufacturing and a whole gamut of other internal departments.

- The regulatory manager in the local country is a key customer as they often have the ultimate responsibility for the artwork of the product in their market.
- The packaging facility is a key customer as they are the next link in the supply-chain to provide the product to the market.
- The quality department is a key customer – at a corporate level to ensure company standards are maintained and each qualified person who has the accountability to make sure the product is correct at batch release.

This broad range of customers also has a broad range of expectations for the service.

How do you measure success?

The broad range of expectations creates a broad range of potential measures, so the key measures and subsequent targets need to be clearly defined in a balanced scorecard to make sure that each stakeholder's needs are suitably addressed. A typical balanced scorecard is discussed further in the supporting processes chapter.

What do you need the service to achieve?

What are your business objectives and therefore what do you need the service to deliver now and in the future? Some typical examples are:

1. An expanding asset model where either existing products are being launched in further territories or new assets are being developed or purchased for existing territories (or any combination of these). This is going to drive increased numbers of artworks and components, further languages, new regulations and regulatory environments. This will drive increased volume. Do you have the processing capacity for this increase and are your processes and systems sustainable and your people capable of coping with this increased complexity?

2. Joint ventures and acquisitions. Are you clear on the rules for the production of artwork and packaging components in an increasingly diverse business model? If you are introducing new operating scenarios, does everyone understand the rules of each "game", and are they sufficiently discrete that an individual will spot that they have to deal with a different operating variant for any specific change?

3. Are you divesting or licensing products? What service may you have to provide to external parties on a temporary or ongoing basis? Will they happily accept your ways of working or insist you comply with theirs? Will they require you to work with different service providers (translation providers, artwork houses, print suppliers)?

4. Do you have products that require highly frequent changes? Some products (for example seasonal vaccines) require very frequent changes to the labelling and artwork often under very short timescales. What capabilities do you need to have to introduce such sequential changes effectively without confusing the organisation or risking high material stock obsolescence?

5. Is your business contracting? How do you ensure you are delivering a cost-effective service that can be down-sized whilst maintaining the levels of quality assurance that your customers and regulators expect?

These pose many questions about the capability you need in your service and how you expect it to evolve over the future. It is good to understand these up-front to ensure that your service capability development is future-proof.

Who "owns" the service?

The cross-functional nature of the service makes ownership difficult. We would suggest that there needs to be a suitable cross-functional governance team with appropriate representation from impacted functions, but also a chair of this group who can act as the overall sponsor and figurehead. Typical governance structures are discussed further in the organisation and supporting processes chapters.

Who is involved in the service?

As mentioned earlier, the service impacts many groups in the organisation. If the goals and measures for the service are clear then the expectations on specific roles and individuals can also be clarified and built into appropriate performance appraisal systems. The service presents a classic 80:20 situation as there are a small number of individuals who are heavily involved in the service (for example artwork operators) whereas there are a large number of people who do small parts of the process as a small part of their jobs (for example local country regulatory managers). This means that the way people are engaged in the service is different. The communication methods, prioritisation setting, performance management and education and training will all be subtly different. Have you considered how you need to adapt each of these for the different audiences in the service? Regardless of which group an individual may sit in, they should be able to identify a member of the governance team as the representative for their function and a place to escalate concerns or issues or seek guidance.

Service statement

In order to answer the above it is good practice to capture the requirements of the service in a service statement of some kind. This may take the form of a service level agreement or any other similar document used in your company. It gives clarity to everyone within and outside the service on what the service is and is not there to do, how success is measured and how the service is expected to grow.

Guiding or underpinning principles

To support the service statement, it is also useful to define a set of guiding or underpinning principles on how the processes and capabilities will operate. These define the "rules of the game" and will help all parties involved in delivering the service when having to make decisions about how to move forward in a particular situation. Figure 3.3 provides some examples of typical service principles.

Examples of Artwork Service Principles
- Single set of global processes
- Competency standards required for all roles
- Capability managed under corporate Quality Management System
- Artwork creation performed in dedicated service centres
- All artwork managed under corporate change control system
- Defined, mandatory, critical control points
- Standard lead times for most changes
- Single, coordinated improvement programme

Figure 3.3 Examples of artwork service principles

They provide value when changes to service, processes or capabilities are being proposed as they can be used as a framework to assess the impact of the change, and if the change will take you directionally towards or away from your desired operating state. In complex cross-functional environments with multiple conflicting goals, a clear set of principles thought out at the beginning of a change journey can prevent many wasteful and painful debates later on.

Service culture

Developing a common service culture across the various teams involved in delivering the overall artwork capability is also a useful means to ensure successful delivery of the service.

It must also be recognised that, in providing a service to a broad group of stakeholders, it is rarely, if ever, possible to please everyone all of the time. An element of good service management not only recognises this, but actively helps to ensure its key stakeholders also recognise this and are involved in collaborative decision-making for key aspects of the service delivery.

It is easy for an external supplier to develop a service culture; after all, it is inherent in the nature of the relationship between the two parties. Not pleasing your customer on an ongoing basis more often than not results in a clearly recognisable termination of the relationship.

When managing internal service functions, the service nature of the relationship between the artwork capability and the rest of the organisation is not as obvious to everyone involved unless it is carefully orchestrated. This requires activity not only on the part of the group providing the service, but also on the part of the customer groups. As with relationships with external providers, it is all too easy for a customer group to abuse the relationship and blame the service provider for all manner of issues. To be successful, the service group and the customer groups should strive to see the relationship as a meeting of equals for mutual benefit, not a master and servant relationship.

You will also recognise that the artwork service relationship, if it is to be successful, will last a considerable period of time. Indeed, if the service is provided by a largely internal team, there is little or no practical opportunity to stop the relationship. Everyone in a long-term relationship will recognise that, for the relationship to be successful, effort needs to be put into it from all parties. Managing an artwork service capability is no different and this effort needs to be budgeted for and the necessary work planned and executed.

4
The core artwork process

As we have discussed in earlier chapters, creating correct artwork is an activity that requires many groups to act together in an orchestrated way to deliver a successful result, on time. The way of ensuring that these people act together in a co-ordinated way is to define a set of processes that everyone adheres too.

Whilst there will always be many ways to reach the same result, and artwork creation is no exception, we will present a high level process here as a basis for discussion. This process is based on experience working with a number of different companies, and if you are involved in artwork processes we are sure you will recognise many elements of it.

Before we go any further though, we would like to re-emphasise that what we present here, and in subsequent chapters, is not something that can just be picked up and used in an organisation. There are many differences in each organisation that will dictate modifications to the process, either at a high level, or in the detail of each step. Furthermore, we make no attempt here to define the details of the individual steps of the process as that is beyond the scope of this book.

We have divided our discussions about artwork-related processes into three distinct areas, in an attempt to make things clearer. The three areas are:

Process area	Description
Core process	The primary activities involved in defining and executing individual artwork changes.
Interfacing processes	Those business processes that interact directly with the core process, and will have an influence on the core process and may be modified as a result of this interaction.
Supporting processes	The business processes that are required to support the core process and other artwork capabilities.

We will deal with each of the process areas in different chapters and will start here with the core process.

Defining the start and end points of the core process

Before defining the process in detail, it is useful to define the inputs and outputs of the process or, put another way, its start and end points.

Typically, the main inputs to the process are:

- A trigger for change.
- Approved regulatory text.
- Other information derived from the company core data sheet.
- Local packaging legislation requirements.
- Packaging component technical specifications.
- Production artwork livery and design guidelines.

There is often a debate when pharmaceutical companies come to defining the start point of the artwork process which relates to the topic of translations. In this context, translations are taken to mean creating local language artwork text from an approved source language text. This situation is common in most pharmaceutical companies where the company will create and manage some form of single language company core data sheet (CCDS), defining things such as product descriptions, prescribing information, master labelling text etc.

Some organisations have a well established process to manage the creation of local language translations and provide them as an input to the artwork process. In other organisations, this translation process is carried out within the existing artwork process. For some companies, this problem will be new as they expand into new markets. Whatever the situation, a conscious decision needs to be taken about where to manage translations moving forward.

The typical outputs of the process include:

- Approved artwork.
- Approved bill of materials.
- Approved samples and first packaging components.
- Archived audit trail with associated evidence.

When an organisation starts to define its artwork process, there is always a debate about the end point of the process. Most people would agree that the process needs to demonstrate that any new or modified artwork makes it on to real packaging components successfully. It is at this point that opinions start to diverge.

Some would argue that it is sufficient to ensure that the first time a new or modified component is successfully quality checked at receipt into a packaging facility, that the process has demonstrated success. Others would argue that the process is not complete until the first packaging batch using new or modified components has been released by the quality department.

A further driver which influences this debate is the increasing need for organisations to report to external regulators when critical safety changes have to be completed. This can lead an organisation to consider extending the process to track the availability of packaged product containing particular new or modified packaging components in local market warehouses, or even the first sale of product. In our experience, organisations normally have other business processes which can track this sort of outcome more effectively and therefore we have not included it in our discussion of the artwork process in this book.

For the purposes of this book, we will assume that the process finishes when the first batch of new or modified packaging components is approved by the quality control group at receipt into the packaging facility.

Triggers for change

There are a number of events within or external to a company which can trigger the need for the development of new or modified artwork. One of the challenges for any organisation is to put in place mechanisms that funnel these triggers to initiate the appropriate artwork creation or changes. Whilst not the subject of this book, we touch further on this topic in the interfacing processes chapter.

Typically, triggers include:

1. **The requirement to change the Company Core Data Sheet (CCDS)**
   Where there is an identification to change information held within the CCDS that is relevant to patient-critical information contained within packaging artwork.

2. **The requirement to change the core regulatory text**
   Where there is an identification to change regulatory core text, either regionally or locally, which does not impact the CCDS.

3. **The requirement to change other information on an artwork**
   Where a requirement is identified to change non-core text, i.e. that which is not on the core text document. Examples of this might include: address changes; trademark changes; barcode changes.

4. The requirement to change the brand image of the product or pack
   Where the brand image of a product or pack is required to be changed, this will drive a change in one or more artworks.

5. The requirement to change the physical design or dimensions of an artwork
   If the physical structure or size of a pack are identified as requiring a change.

High level process steps

At its highest level, creating artwork is no more complex than defining what is required, creating a work product such as an artwork and then verifying that this output meets the requirement initially defined. This is a very familiar process to anyone involved in quality systems.

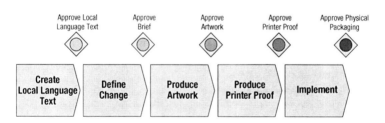

Figure 4.1 Level 1 core process and critical control points

For the purposes of this book, we have defined a high level process consisting of five fundamental, or level 1 steps. Furthermore, each step culminates in a critical control point to ensure that it has been performed completely and correctly before moving on to the next step. The five steps are represented diagrammatically in Figure 4.1 and described in Figure 4.2.

Level 1 Process Step	Short Description
1. Create Local Language Text	Create and approve local language source text document(s) for each of the packaging components to be created or modified.
2. Define Change	Define exactly what is required to be created or modified as part of this change.
3. Produce Artwork	Produce a new or revised artwork that complies to the requirements defined in the define change step.
4. Produce Printer Proof	Produce a modified artwork file that can be used directly in the packaging component printing process. This file differs from the artwork produced in step 3 in that it is modified to include all features that will allow it to be successfully printed via a specific printing route. *It is possible to eliminate this step through the use of a print ready process.*
5. Implement	Ensuring that, at minimum, the first time a new or modified artwork is used to create packaging components for use in the manufacture of real product, that they are correct.

Figure 4.2 Level 1 core artwork process description

Critical control points

It is worth pausing at this point and briefly discussing process critical control points. Given that this process produces information that, if incorrect, can have a significant and potentially fatal impact on patients, it is critical that there are defined control points in the process to ensure that the quality of the output of the process is to the highest standard practically achievable.

To achieve this, companies have found it useful to define critical control points in the artwork process to ensure that all necessary tasks have been completed to a high quality before moving to the next phase. Each control point would normally include a quality check for accuracy as well as a formal approval by key individuals to proceed. In addition, some control points will provide an approval of a master document which will form a part of a master record source for GxP information.

As we will discuss later in more detail, we would recommend that the execution of key control point activities is supported by the use of checklists to ensure that all activities are consistently and completely performed.

Core process detailed discussion
Step 1 – Creating local language text

The overall process begins with the creation of local language text. As has been discussed earlier, this step may be considered by some organisations to be part of another process which provides the source text as an input to the artwork process, but for completeness we have included it here.

For a new packaging component, this step normally takes as its input some form of product company core data sheet (CCDS). The CCDS documents constitute the master reference for the company and would normally include, amongst many other things, master labelling text in a single language. There is normally a rigorous review and approval process within a company to ensure that this master reference information is correct. The scope of the preparation and approval of CCDS documents is outside the scope of this book.

Normally driven by the regulatory function, this process takes the CCDS labelling text and creates local language labelling text documents which define the text to be included on each packaging component. Clearly, this step includes the translation of text from one language to another, a process which requires appropriately skilled individuals to perform correctly without changing the context or intended meaning of the text.

For an existing packaging component, the bulk of the local language labelling text will already be available from a previous artwork change. All that is required in this case is to identify any changes that are required to it. Clearly, there needs to be a decision here as to whether the local language text document must also be updated.

The process of engagement with the relevant external regulatory agency to agree local text varies with country or region, but can be broken down into four main groupings. The artwork process needs to recognise that reaching this agreement often involves much iteration.

European centralised procedure
A centralised and formal European Medicines Agency (EMA) process for the central registration of pharmaceutical product for all European Union Member States.

European mutual recognition
A centralised and formal EMA process for the registration of pharmaceutical products where one member state takes the lead role for registration of the product and other member states progress from that recommendation.

National product registration
The registration process follows the requirements of the individual regulatory authority. This varies by the individual regulatory authority and is mandated for registration of pharmaceutical products in the relevant country.

Other national arrangements
Varying national processes whereby some individual regulatory authorities accept the use of products from other countries without any change to the originating country packaging or artwork design.

The nuances and complexities of these processes are too great to cover in this book, and are subject to ongoing development and change. We guide you to the EMA website for further information on European centralised procedure and mutual recognition and individual authorities' websites for national processes.

The provision of appropriately skilled individuals to undertake translations provides some challenges. Typically there are four solutions that a company could use:

- Using the local market company staff that are fluent in that local language.
- Using central staff who are fluent in the local language in question.
- Using a local translation agency residing in the local market.
- Using a centralised regional or global translation agency.

Each of the above options provides benefits and challenges. An error in the translation will result in an error in the final packaging component and therefore needs to be appropriately controlled. Some key things to consider in this management process include:

- Ensuring the meaning of the translation is correct in the local language.
- Ensuring the accuracy of translations versus the core data sheet. Undertaking back-translation is one technique that can help here.
- Ensuring the competency of internal staff and external service providers who undertake translations. It is often considered necessary and prudent to use only native speakers with medical training to perform translations into their native language. This can help to address the first point.
- Ensuring that requested changes are valid and not just personal opinion or preference.
- Ensuring consistency of translations across different products in a local market.
- Avoiding wasted effort by translating the same phrases multiple times for each instance where they are used.

We discuss translation further in the technology chapter, as IT tools can help to address some of these considerations.

The text created by this process step may not be the entirety of the text to be included on any one packaging component, but it will normally contain all the medically critical information as agreed with the relevant external regulatory authority.

As the final activity in this process step, there would normally be a formal critical control point to approve the local text document(s) prior to commencing the next step of the process.

Step 2 – Define change

As we know, artwork is the culmination of information from many different sources, both internal and external to the organisation. Furthermore, in many cases, the same information gets repeated several times across one or more artworks.

When a change is triggered, there is a temptation for organisations to jump straight into designing artwork, using the artwork itself as the vehicle to gather the necessary source information. We would not recommend this approach for a number of reasons, the principal one being that an artwork is a very poor vehicle to capture many elements of source data in a clear and concise way. A number of organisations that we have observed which have taken this approach tend to exhibit a number of undesirable results, examples of which include:

- Multiple cycles around the artwork development process, often leading to…
- Very long lead times in the creation of artwork.
- Frequent omissions and errors.
- Stress for the individuals carrying out the process as they attempt to remember all requirements.

It is for the above reasons that we would recommend that the artwork development starts with the creation of a clear requirement that captures all the relevant source information for a new or revised artwork in a structured way. This requirement documentation is often referred to as an artwork brief, or brief for short, and this is the term we will use for the rest of the book. In reality, the brief is a collection of information and source documents that paint a complete and unambiguous picture of the required artwork.

The brief clearly provides a complete and concise source of information on which to base any creation or modification to an artwork. It also serves another very important purpose, that of the source document against which a new or modified artwork can be compared when checking and approving it. The provision of clear and precise information in the form of a brief can be a key contributor to the reduction in artwork-related errors.

As the brief is the complete definition of the required change that will be referenced when approving the artwork, it should be formally approved by an appropriate group of stakeholders. This approval forms the critical control point at the end of this step of the process. Once approved, the brief should be archived to form part of the audit history.

We will now discuss some of the key activities normally involved in creating the brief.

What is it that is changing?
There needs to be a concise statement about what it is that is changing and why, written in a way that all stakeholders will be able to understand.

Impact assessment – which artworks?
It is easy to focus on the requirements for individual artworks when considering the whole artwork process, but this would be a mistake.

If we recall the earlier discussion on triggers for the process, we can see immediately that many of the triggers can give rise to the need to change several product packs and, more often than not, several artworks. A key step in defining the change requirement resulting from a trigger is to identify all the potentially impacted product packs, components and associated artworks. Indeed, one of the main causes of error in artwork is caused by getting this step wrong and completely omitting to change one or more impacted artworks. The result of this assessment would often be captured in a draft bill of materials, detailing all the components (and therefore artworks) that will be included in the new or modified product packs, indicating which of these components is to be changed and which are to remain the same.

One area to pay particular attention to in the impact assessment is the use of one component in multiple packs. The impact of the change on such components needs to be assessed for all products in which the component is used. A variant of this situation is the use of a pack in several markets, particularly where the market accepts another country's products in certain circumstances. Again, assessment of the change for wherever the product is used needs to be assessed.

A check should also be made at this time to ensure that other changes are not in progress on the impacted components. If they are, then decisions need to be made about how to progress.

Furthermore, it is important to note that artwork changes are often only one of many organisational impacts which result from an individual trigger. At minimum, an artwork change will result in the need for several supply-chain related actions to be carried out to ensure that the new or modified artwork appears on packaging components in a timely and co-ordinated manner.

In some organisations there may be an existing and robust mechanism to identify the impact of a trigger, in others there may not.

The important thing here is that, as part of the design of a robust artwork process, decisions need to be made as to how and where this impact assessment is carried out and to incorporate the appropriate resulting activities into the artwork process.

Impact assessment – content and design

Once all of the impacted packaging components have been identified, it is time to assess the impact of the change on the design of individual artworks. To illustrate this level of impact assessment, we will look at a couple of examples at different ends of the complexity spectrum.

Let's start with a straightforward change, which may be as simple as specifying a simple text change to an existing artwork, for example an address change. In this situation, there is very little unknown about making the change: the external regulator may only need to be informed about the change, rather than being required to approve the change; the same physical packaging component will be used; the same artwork layout, content and colours will be used, with the exception of the small area of changed text; the same printer will be used; the same supply-chain route will be used to supply the components; the same packaging facility will be used to pack the product. Therefore, given the very confined nature of the change, it seems reasonable that it is not necessary to involve a significant number of stakeholders in assessing the impact of the change. Put another way, the risk of the change causing issues is very low and therefore it is not necessary to involve a significant number of stakeholders.

Indeed, in this sort of situation, as far as the change to the artwork is concerned, it may only be necessary to involve the regulatory, local market representative and perhaps legal team in defining and agreeing the new address text. From a technical perspective, there are no risks of knock-on impacts in making the change, so it is not necessary to involve the packaging technology group, the printers or the packaging site in defining the change.

As is common to all changes, the supply-chain planning team should be involved in assessing the impact of the change to ensure that the plan of activities for the change aligns with other supply-chain activities. As was pointed out earlier, even simple changes like this need to be co-ordinated with other supply-chain activities to ensure that the new components are delivered to the packaging site on time, old components are run-out in an efficient manner and the appropriate updates to specifications, ERP systems etc happen in a co-ordinated fashion.

Planning consultation should also occur with all impacted functions to ensure that resources are aligned to deliver the change in the required turnaround time. To avoid having to consult many stakeholders about the timing of every change, it is normal to adopt common lead times for each process step.

The result of the artwork-related impact assessment in this case will be a new address text, clear instruction on how and where that new address text is to be applied to the artwork and a plan of action, including target dates for all key activities.

Having got this far, the only thing left to do in this process step is to pull all the information defining the change into one place – the brief – and get the right stakeholders to sign it off as correct. In this case, the signatories might be the local regulatory representative, local market representative and a supply-chain representative.

Once approved through formal signature, there are a couple of essential things to complete prior to moving on to the next stage of the process. Firstly, the documentation associated with preparing the brief needs to be appropriately archived or destroyed in line with the company's information retention policy. Secondly, the right people need to be informed that the brief has been approved and is on the way through the artwork change process. This last activity is important to ensure that all the necessary resources are aligned and made available to ensure that the change happens according to the agreed plan.

Now let's consider a more complex change. Let's say this change requires a new multi-market artwork to be put on a new physical packaging component, utilising a complex new printing technology at a new printer and packaging facility. Clearly, the extent of the change and the risks involved are significantly higher than in the previous example. However, the principal difference in the approach for the more complex change is the involvement of a broader group of stakeholders in assessing the impact of the change. In this case, multiple local country regulatory and marketing representatives will need to be involved due to the shared nature of the pack in question. Furthermore, due to the technical risks involved, it would be prudent to involve the packaging technology group and the printer. Given the new packaging site involved, it would be prudent to involve them as well to ensure that the resulting component can be packed effectively on their packing lines.

Artwork mark-up

One useful tool that we did not touch on explicitly in the above discussion was that of how to define exactly what changes need to be made to an existing artwork, such that the description is unambiguous to those carrying out or approving the change. This is where the artwork mark-up comes in. The concept is simple: in order to explain what changes need to be made to an artwork (essentially a two-dimensional picture), we need to describe where a change is to be applied and what the change is. To do this most effectively, physically marking up an existing artwork is the best way forward. Figure 4.3 gives a simple example of this.

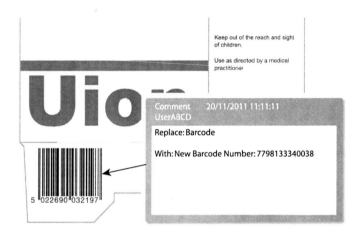

Figure 4.3 Example artwork mark-up

Step 3 – Produce artwork

Now we have clearly defined what is required for a new or modified artwork, it is time to create the artwork. This step of the process consists of creating the artwork, checking that it meets the requirements of the brief, approving it and then archiving it.

Before we go on to talk about what this step involves in more detail, we will pause to discuss what it should not be. The brief defines what is required of the new or changed artwork and was approved by all the appropriate people. The artwork creation stage of the process is not a continuation of the development of the requirement for the change; it is the execution and verification of the change. Therefore, if it is realised during the artwork production phase that there is a material change required to the brief, then we would recommend that the process should loop back to the brief stage to clarify the requirement.

Artwork creation

Creation of artwork should be performed by trained artwork operators, using controlled artwork creation tools. Some companies use in-house staff to perform this activity, others use external artwork creation companies. We will discuss the potential use of external companies for this stage of the process in the outsourcing chapter.

Recently, there have been developments in the area of automated artwork creation. This method for the creation of artwork has not been adopted in the mainstream for pharmaceutical artwork creation at the time of writing this book. We therefore discuss it further in the chapter on future developments.

If you put yourself in the position of an artwork operator for a moment, you will quickly realise that the structure and content of the brief is key to being able to perform your job efficiently, effectively and without error. Fundamentally, the creation of artwork is all about laying out information, both text and graphics, on to a pre-defined background, the blank packaging component. This immediately starts to help us define some aspects of the brief that should be defined in order to make the process of artwork creation effective.

Firstly, the text and other critical information, such as logos, need to be identified clearly and unambiguously in the brief. For this reason, creating a brief in a standard format will help the artwork operator find information efficiently. This information may be laid out on a form, or be provided in source documents which have a standard format.

Many errors occur when information is transposed, or re-typed, by operators. Therefore, wherever possible, processes should be designed to allow artwork operators and others to cut-and-paste information from one electronic document to another. Amongst other things, this has implications on the format of information source documents; as an example, source information in the form of PDF files should have selectable text feature enabled.

Secondly, for the artwork operator, it needs to be clear where each element of information needs to be put on the artwork. It is here that the practices of companies vary widely. At one end of the scale, there are companies who leave layout almost entirely to the artwork operator, basing their judgement on experience and previous examples of similar or related artwork. At the other end of the scale are the companies who highly structure their artwork using layout templates which define where each element of text and graphics is to be placed. In this instance, the artwork operator is essentially executing a very highly defined instruction of exactly what to do. The latter model is more common in fast-moving consumer goods companies, where the frequency of change is high and the need to maintain consistent brand image across multiple products and markets is critical. A middle ground involves the use of corporate brand guidelines and/or model artwork designs to define the look and feel of each type of artwork, which the artwork operator then uses as guidance when creating each individual artwork.

Artwork document structure

It is important to recognise that the way an artwork is constructed is important. An artwork that is well constructed will:

- Aid in a structured approach to creating the artwork.
- Simplify and assist the review process.
- Be acceptable to the various electronic tools that may be used to assist the artwork process.
- Ensure the artwork requires little or no manipulation in the pre-press process and therefore minimises non value-added work and the opportunity to introduce errors.

To this end, there should also be technical guidelines that define aspects such as how an artwork will be constructed and layered.

Artwork review and approval

Once created, there needs to be some form of review and approval of the artwork. If we assume that the brief that was described earlier has been created and approved, then the review and approval of the artwork needs to verify two principal things:

- That all changes defined in the brief have been made, and...
- That no other change has been inadvertently made.

There are several different aspects of an artwork that need to be checked during the review step. These can be categorised as:

- Text content.
- Graphical content.
- Technical content.

Text content review
The text verification ensures three aspects of the text in any given artwork. Firstly, it is important to verify that text has been transcribed from the brief or other source documents. It must be verified that all text has been transcribed on to all relevant faces of the artwork and that none of it has been inadvertently hidden. Secondly, checks should be made to ensure that critical information such as product name, strength, dosage etc is correct. Finally it is important to check that the layout of the text has not altered its meaning.

Graphical content review
The graphical checks verify that all the graphical elements of the artwork are as required by the brief. Graphical elements may include logos, branding images, colours etc.

Technical content review

The technical review ensures that all other aspects of the artwork are correct. This will include checks to ensure that items such as dimensions, barcodes, varnish layers etc are correct.

The activity of proofreading and checking artwork is an activity which has all the attributes that would lead to human error and where such an error can lead to significant business risk. Let us take the example of an operator having to compare an artwork with a regulatory source text document to ensure the source text has been transcribed on to the artwork without error. In this case, the operator has to read every word, letter and symbol in both documents to ensure that they are the same. This may be fine for a small label, but for a multi-page, multilingual leaflet, you can imagine that things can get challenging very quickly. Indeed it can take up to a day for a skilled operator to proofread such a long leaflet.

To make matters worse, at least in the context of proofreading, the human brain is excellent at filling in gaps in information and correcting mistakes in information so that it can see meaning very quickly. As an example, try to read the following:

Cna yuo raed tihs? 55 plepoe out of 100 can.

i cdnuolt blveiee taht I cluod uesdnatnrd waht I was rdanieg. The phaonmneal pweor of the hmuan mnid, aoccdrnig to a rscheearch at Cmabrigde Uinervtisy, it dseno't mtaetr in waht oerdr the ltteres in a wrod are, the olny iproamtnt tihng is taht the frsit and lsat ltteer be in the rghi t pclae. The rset can be a taotl mses and you can sitll raed it whotuit a pboerlm. Tihs is bcuseae the huamn mnid deos not raed ervey lteter by istlef, but the wrod as a wlohe.

Many people have little difficulty reading this example; however it does have to be said that some people will find the text unintelligible. Indeed, on first reading, some readers will not notice any issues at all with many of the words. If you look closely at the text, you will see that the middle letters in all the words are actually scrambled, with only the first and the last letters of each word being in the correct place. This has profound implications on proofreading, particularly when comparing text. We effectively see what we want to see, so if I am looking for the word "animal" from the source text and I see "aniaml" in the artwork, I am likely not to notice the error at all as my subconscious mind automatically makes the correction.

For these reasons, people doing manual proofreading activity need to be taught to compare information in a way that attempts to stop the human mind making these subconscious corrections. One way to achieve this is for a proofreader to read the text backwards and character by character. In this way, the human brain has little opportunity to seek meaning from the information in the normal way and the probability of errors being missed is reduced.

An important point worth noting here is that errors occur not only in areas of an artwork that were supposed to be changed, but also in areas of the artwork that should have remained unaltered; for example, information may get hidden by new content such as tables or illustrations. The cause of these undesired changes is, more often than not, human error. It is for this reason that any verification process needs to look not only at what was supposed to be changed, but must also verify that nothing else was inadvertently changed by mistake. Furthermore, it will be obvious to you from this discussion that these checks need to be done each and every time an artwork has been, or could have been, modified.

Typically, the artwork review process involves three groups of people:

- The artwork operator will check their own work.
- A skilled proofreader will check that the artwork text, graphical and technical content is correct according to the brief.

- Stakeholders from the business will verify that key information that they are responsible for is correctly interpreted on the artwork and that the overall result meets their intended need.

This last check is an important one as there are many opportunities for the desired effect of a change to be misinterpreted in some way in the final artwork. Let's examine this a little further.

Firstly, it is worth remembering that people think about their requirements in different ways. Let's keep it simple and consider that some people are visual and some are audio. The former can explain what they want in pictures, whilst the latter are much more comfortable writing descriptions of what they want. Elements of the way the brief is structured should suit both, but sometimes people will end up having to explain what they want in an uncomfortable format. Furthermore, many people find it really difficult to visualise the impact of their desired change on the resulting artwork. If you have ever been involved in designing a house, factory or office, you will probably have recognised at some point in the process that some people struggle to look at a set of instructions or a plan and visualise the end result. When the end result is put in front of them they say things like "I never thought it was going to look like that," or "You never told me that was going to be there."

Secondly, we must remember that many stakeholders in a global pharmaceutical company will be interacting with the artwork process in their second, or even third language. This raises many opportunities for ambiguity in the definition or interpretation of requirements, all leading to potential artwork errors if not reviewed appropriately.

It is therefore important to give key stakeholders the opportunity to look at the end result of the artwork change and review it with their own intended change in mind. Only in this way can an organisation have confidence that the artwork is as the stakeholders intended it to be.

More often than not when reviewing a company's artwork capability, we find that one of the principal reasons that the review and approval steps fail is that people are not clear what they should be checking during these steps because the roles and responsibilities definition for the process is not detailed enough.

With little or no guidance people will, more often than not, do what they think is appropriate in the interests of the business and the patient. However, in our experience this will rarely result in a robust and consistent process that achieves the intention. You will probably have seen the situation in this or other business processes: some people check everything they feel they possibly can, whilst others check the bare minimum which relates only to their narrow area of competence. The resulting artwork has some elements that have been checked by many people and other elements that have not been checked at all. This results in a significant risk that errors will get through the process and result in potential patient harm and/or recall.

If individuals believe they are being asked to check all information, much of which they have no way of knowing is correct, they will set about asking others. This slows the process down and, perhaps more importantly, quickly leads individuals to feel like they are wasting their time. Over time, this leads to dissatisfaction that can result in the degradation and ultimate collapse of the effectiveness of the overall process, something which needs to be avoided at all cost.

It is for these reasons that we would strongly recommend careful design of the checking activity to ensure that an individual is asked to check only the information of which they clearly have authoritative knowledge. We would recommend the use of role-specific checklists, described in more detail later, to capture the specific items of information that an individual needs to check.

Furthermore, we must also remember that human beings, by their nature, make mistakes and omit to do things sometimes. It is for this reason that we would recommend designing manual checks to ensure that two people check the same information – a practice that is well embedded in GMP. Where validated technology is used to assist in the verification processes, we would suggest that the need for a second human check can often be omitted.

Internal artwork approval

Having reviewed the artwork, the next – and some would argue the most important – critical control point arrives: the formal approval of the artwork. At this point, we would recommend that all those stakeholders who approved the brief should formally approve the artwork, recording their approval through signature. This approved artwork needs to be kept as a prime document within the company's quality management system.

External artwork approval

Having internally approved the artwork, it may be necessary to get the artwork approved by an external regulator. The process of formal submission to an external regulator, negotiating and agreeing changes with them is normally the subject of a completely separate business process and is outside the scope of this book. The important thing for the artwork process is that the submission happens and that, at some point, a formal response is received indicating that the artwork is approved, or that it was rejected and needs changes. If the artwork requires change then we would recommend that it should be recycled through the briefing stage again to ensure that the change is documented and agreed formally. If approved by the external regulator, then the artwork can proceed to the next stage of the process.

Any interface between two processes, like the one just described for external regulator review, presents an opportunity for error. We are aware of a number of occasions where companies have not had a formally defined and managed interface in place at this stage which resulted in either artworks that had not been approved by the external regulator getting out into the market, or where changes required by a regulator were not included in the artwork. It is critical that there is some form of unambiguous interface between two such critical processes. A formal interface can be paper based, perhaps requiring an internal memo to document the receipt of the external regulator's approval. Alternatively, it could be electronic, perhaps a workflow tool used to drive the artwork process, preventing passing this stage of the process unless electronically signed by the regulatory department representative.

Having received final approval from the external regulator, the artwork can be archived and appropriate stakeholders in the organisation can be informed. Depending on the specific business, this step may trigger related processes such as activating bills of materials (BOM) in an ERP system.

Step 4 – Produce printer proof

In most traditional artwork processes, the next stage of activity is typically referred to as pre-press. It involves modifying the electronic artwork file into a format which can be used to create print press plates or films and then verifying that the resulting electronic printer proof is correct.

Pre-press activity typically involves the following steps:

- Check that the way in which the artwork has been created is consistent with the printing process that is to be used. This involves verifying matters such as the colours have been defined correctly.

- Setting trapping and adjusting bleeds to ensure that the artwork will print correctly given the specific characteristics of the printing press and inks that will be used.

- Step and repeat the artwork image across the face of the press plate as a press plate will normally contain many images of the artwork.
- Colour separate the artwork into its constituent single colour images, one associated with each press plate that will be used with a single ink on the printing press.

Clearly, all the activities we have described above need to be performed in order to have a printable artwork. You have also probably realised that, given that all of these processes manipulate the artwork in some way, there is a significant opportunity to introduce errors at this stage. Therefore, either the process that carries out these activities needs to be automated and validated, or the result of the activity needs to be checked for correctness. The reality for most current processes that we encounter is that some form of checking is done at one or more stages.

The ultimate check is to have a printer create a sample run of artworks from the print press that will be used for the production of the final components. This is often termed a wet proof. The printer would physically send this wet proof to the customer for final approval prior to releasing any printed components. The use of wet proofs is becoming less popular for a number of reasons. Principal in the downsides of the wet proof is that it introduces a significant lead time into the creation and approval of artworks. Secondly, with increasing automation being used by printers, there is a much higher degree of confidence that the latter stages of the pre-press process can be carried out with a very low probability of error.

Most companies have taken a hybrid route and opted to use an electronic printer proof rather than a wet proof. This electronic printer proof file is the result of many, but not all, stages of the printer's pre-press process and provides an opportunity to verify that the artwork has not been adversely impacted by the majority of the pre-press activities. Here, the printer would send the electronic file to the customer for approval prior to printing any components. This clearly has the advantage of speed, in that no real print run is required and electronic files can be transmitted almost instantaneously.

Printer proof checking
Once returned from the printer, the printer proof, in whatever form it takes, needs to be checked against the previously approved artwork to ensure no unacceptable changes have been made. Because there are not supposed to be any material changes at this point, the type of checking techniques used are somewhat different. Here the focus needs to be on two principal things:

- Verifying that a gross error has not been made, such as using the wrong artwork.
- Verifying that the printer proof is graphically the same as the approved artwork "pixel for pixel".

Because we have an approved artwork to compare the printer proof against, there is no need to involve all of the stakeholders again in the printer proof review and approval stages. Normally, only a proofreader would be involved in this step.

In a manual world, transparencies and similar tools would be used to compare the printer proof graphically with the approved artwork. These techniques can be effective if carried out by a highly trained operator with the right tools in the right environment; however, they are always subject to operator error. We will discuss the opportunities available to use electronic tools to assist this step of the process in the technology chapter.

Printer proof approval
Approval of the printer proof, in whatever form, represents another critical control point in the process and, as such, the printer proof should be approved by the signature of one or more competent people and archived.

Typically, the proofreader who performed the checks described above would sign their approval of the printer proof. Typically, a second person would also check critical information on the artwork to ensure that no gross errors had been made and again sign their approval.

Elimination of the electronic printer proof
Whilst pre-press activity is always required in some form or another, the need to have an electronic printer proof created and approved is not.

For example, if there are adequate and complete checks of the printed packaging components prior to use, it can be argued that this step of the activity is only reducing the business risk of mistakes being discovered later and causing significant corrective action delays.

An alternative argument for the removal of the electronic printer proof involves the use of recent developments in technology. Artwork can be constructed in such a way as to require no change to the image that will be reproduced on the packaging component during the pre-press process. Furthermore, the increasing use of automated electronic tools to carry the remaining pre-press activity also serves to reduce further the probability of errors occurring. With this in place, some companies are comfortable to eliminate the need for the production and approval of the electronic printer proof.

Some packaging components are printed on electronic printing presses which require no press plates or films to be created. If artworks are produced and approved in a format which can be directly input to these electronic printing presses, then the need for the production and approval of the electronic printer proof can be eliminated. At the time of writing this book, this type of printing technology is typically only economic for very small print quantities and has some other limitations associated with it. It is therefore not generally used for printing packaging components to any great extent in the pharmaceutical industry.

Step 5 – Implementation

As we discussed earlier, for the purposes of this book we will take the end point of the process to be the approval of a printed packaging component at the packaging facility.

Component ordering co-ordination
The first activity in this process step involves ordering a batch of printed packaging components. The issue here is to ensure that the business processes involved in this step ensure that the right component gets ordered in the right version. Given that we have just discussed creating a new or revised version of a packaging component, the sequencing of all the activities required to ensure that the right components are ordered is not simple, and many are outside the scope of the core artwork process.

One very straightforward issue that needs to be resolved in the ordering process is how the different versions of components are uniquely referenced by both the customer and the printer. This will have a knock-on impact on the artwork process in that it must be ensured that the relevant component codes and revisions are visible on the printed component to ensure that the correct component is being processed.

Furthermore, the component numbering system should ensure that different revisions of a component cannot be confused.

Component printing
Once ordered, the printer will produce printed components and carry out some form of internal quality check. It is beyond the scope of this book to discuss the processes involved in the printing of the components and it is clear that all printed component suppliers would normally be managed under a formal supplier quality management system.

The internal quality check that the print supplier carries out can vary widely. It is important to understand what they do in each specific case and build the overall quality checking regime accordingly. For example, if a printer has a fully automated and validated vision system that 100% checks every component as it is printed, it may not be considered necessary for the customer to repeat a 100% graphical check of a sample of components. However, if the printer's checking is limited to a few manual checks, then a more thorough customer check may be considered necessary. Unfortunately, for a number of reasons, many pharmaceutical companies take a blanket approach to the internal checking of printed packaging components, assuming very little printer quality checking is being carried out.

Having internally approved the printed components, the printer will then dispatch them to the customer.

On receipt, printed components are normally placed into a quarantine area awaiting quality control inspection.

Assuming that a full inspection is required given the printer's quality control activities, the inspection at this stage would be very similar to the checks made of the printer proof:

- Verifying that the right component is being checked.
- Graphical comparison of the component to an approved artwork image.

In addition to these checks, it would be normal for other specification checks to be carried out such as:

- Materials.
- Component size.
- Folds, creases and embossing.
- Component materials and surface finishes.

- Correct colours, free from smudging.
- Inks dry and resistant to smudging with water.
- Barcodes readable, correct specification, correct information content, acceptable print quality.
- Braille correct and acceptable quality.

It would be normal practice to take a set of samples from the printed component batch and perform these checks on each one. Given that several components will need to be checked for each batch of packaging components, there is a good case for inspection technology to assist this process, a topic which we discuss in further in the technology chapter.

Once verified as correct, the printed components can be approved and released for use in the packaging process and the relevant record of the quality inspection and approval can be retained.

Management of redundant components
There is an area of co-ordination that is worth discussing further at this point as it has created a number of recalls in the past. The management of redundant or superseded components is critical to ensure that they do not get inadvertently used in a packaging batch. The packaging operation would normally require formal processes to identify redundant packaging components, quarantine them and ensure their safe destruction.

A similar issue occurs where a printer creates print plates that are used to print multiple batches of components. There should be a management system in place to ensure that redundant print plates are removed from the normal material flow in a timely manner to prevent their accidental use in production. A slight complication that needs to be considered is sometimes seen in tender business. At the end of a tender's life there may be a requirement to package an additional batch of product after the components have already been updated for a new tender requirement.

Capturing the process

Now that we have discussed the core business process in some depth, we will take the opportunity to discuss a little of how an organisation might go about capturing the process. If you are very familiar with this topic, feel free to skip to the next chapter.

There are several objectives in defining and capturing the business process:

- Ensuring everyone knows how to play their role in the process.
- Ensuring critical work is done in a complete and standard way.
- Ensuring there is a "corporate knowledge" capture on which future improvement can be based.

There are various tools that may be employed to capture the process. There are many different specific ways to describe processes and each has its merits. For the purposes of this discussion, we will use a framework based on levels of detail as follows:

- Level 1 – Policy.
- Level 2 – Standard operating procedures, work instructions and checklists.
- Level 3 – Guidelines and work aids.

The level 1 policy documentation would normally lay out the highest level of requirement for the management of artwork creation and change in an organisation and is normally managed under the company quality management system. It would not normally define how artwork management would be done, just that it needs to be done. In this way, different parts of the organisation are free to meet the high level requirements of the policy in different ways that suit them. Clearly, given that the artwork process is, by its nature, often global and cross-functional, it is worth noting that it may not be considered desirable for different parts of the organisation to have different processes to meet the requirements of the policy.

The level 2 standard operating procedures (SOPs), work instructions (WIs) and checklists are the core documents that define the activities which people have to perform in order to carry out the process and, again, would normally be managed under the company quality management system. SOPs describe the process as a sequence of steps, defining what needs to be done and who is responsible for doing each step.

SOPs are normally written at a relatively high level, providing the "what" and "who", but not much of the "how". They are written at this level to ensure that there is a clear understanding of the end-to-end process at an appropriate level for all those involved. When it comes to the specifics of "how" a particular individual will carry out a particular step in the SOP, this is where WIs, checklists, guidelines and work aids come in.

WIs are designed, as the name suggests, to guide an operator through how to perform a particular step in a SOP. They need to be written in co-ordination with checklists which we will describe next. It is normal to focus WIs on complex activities which need commentary to explain clearly how they need to be done and would often be supported by a corresponding checklist. An example of a step in the process which would normally require a detailed WI would be the proofreading step. Where activities are relatively self-explanatory, a checklist alone will normally suffice.

One of the primary issues that we see with SOPs, and to a lesser extent with WIs, is that people, once trained in an SOP or WI, rarely refer back to it. This means that it is very easy for individuals to omit critical activities or develop their own ways of working, often without necessarily realising it. To a certain extent, periodic refresher training can help this situation, but often this is not very effective at resolving the issue. It is for this reason that we would recommend the development of comprehensive checklists to be used in everyday activity by people performing the process. The checklists should capture the essential activities that need to be performed in each step of the process and, depending on the criticality of the step of the process being performed, consideration should be given to them being signed by the operator at completion of the step and retained in the audit trail of the change. Checklists provide a powerful tool for an organisation to ensure work is done in a standard way every time. Furthermore, they capture a level of detail which provides an excellent basis on which to define improvements to the process, in a form that is highly likely to be enacted by the operating staff.

The level 3 guidelines and work aids define a level of instruction or advice which may be helpful to operators, but is not necessarily mandatory. For this reason, some organisations do not manage this level of documentation under their formal quality management system, whilst other organisations do. Guidelines and work aids might instruct the operator on the way to use IT systems and tools for particular activities, or they may provide guidance in decision-making during certain steps of the process. Whatever they are used for, we would always recommend that they are considered an integral part of the overall process definition documentation set and be formally managed as such, even if they are outside of the company's formal quality management system.

5
Interfacing processes

In this short chapter we will touch on the topic of the processes which interface with the core artwork process. Figure 5.1 identifies many of these typical interface processes and, for each one, highlights what the nature of the interface is.

The first point to note here is that the artwork process does not operate in isolation. It is a process which relies on information and activity in many other processes in order to operate successfully. Furthermore, some of these processes are owned and operated by organisations external to the company who owns the core process.

The design of the artwork process must clearly take account of each of these interfacing processes. For each process it should be clear at which point the interface(s) occur, what information is interchanged between the processes and in what format.

When designing the artwork process, it is highly unlikely that all of the interfacing processes will provide exactly the right information in the ideal format to support the new artwork process. Consequently, analysis will have to be done in each case as to the best way forward. In some cases it will be necessary to modify the interfacing process to meet the ideal needs of the artwork process. In other cases it will be necessary to modify the design of the artwork process to accommodate the constraints of the interfacing process. In many

Typical Interfacing Process	Typical Nature of Interface
Change control process	• Triggers and information to drive the artwork process
Production planning	• Planning information
ERP data management process	• Coordination of the management of information in the ERP system(s)
Physical packaging development process	• Triggers and information to drive the artwork process
Company core datasheet development	• Triggers and information to drive the artwork process
Local and regional labelling information development	• Triggers and text to be put on the artworks
Product code management	• Product codes to be placed on artworks
Component code management	• Component codes to be placed on artworks
Brand design	• Information to help design artworks
Logo management	• Information to help design artworks
Trademark management	• Current trademarks to be placed on the artworks
Product registration management	• Registration codes and addresses to be placed on the artworks
3rd party supplier and partner processes	• Triggers and information to drive the artwork process
Product launch	• Triggers and information to drive the artwork process
Supply chain engineering change project management	• Triggers and information to drive the artwork process
Sales and Operations planning	• Visibility of upcoming demand on the artwork capability
Supply chain design	• Visibility of upcoming demand on the artwork capability

Figure 5.1 Typical interfacing processes

cases a compromise solution will result. In some cases it may be necessary to phase the implementation of the new process, initially implementing a less optimal solution which can later be optimised when the corresponding interfacing process can be modified.

It can be helpful when defining the nature of the interface to consider these processes in a number of different groups.

Data provision processes
Those processes which provide specific information to be included in the artwork. The primary consideration here is to ensure that the information is provided in an unambiguous, complete and accurate format and that the artwork process verifies its correct use in some way.

The product code management process would be an example of this type of interface process.

Guidance processes
Those processes which provide information to guide the production of artwork. The primary consideration here is to ensure that individuals are appropriately trained, competent and aware of the guidance to use it successfully.

The brand design process would typically be an example of this type of interface process.

Management processes
The artwork process typically needs to maintain alignment with activities in several other business processes.

The production planning process would be an example of this type of interface process.

6
Supporting processes

The core process described in the previous chapter defines how individual artwork changes will be carried out. Whilst this is absolutely critical to the success of the artwork capability in an organisation, it is not sufficient in itself to provide a complete capability. A number of support processes need to be in place to achieve this. Figure 6.1 outlines a support process model that we will use in this book. When all these processes are in place, an effective and sustainable artwork capability is more likely.

Many organisations will find that they already have one or more of these supporting processes in place that can be adapted or extended in scope to include the necessary artwork process areas. In many instances, this approach is to be recommended, as the artwork capability does not necessarily need its own unique iteration of a supporting process.

There are a number of questions that need to be considered in making the choice about incorporating artwork into an existing supporting process or creating a separate artwork-specific iteration. These include:

- Does a robust supporting process already exist elsewhere in the organisation which has a close fit to the supporting process requirements for artwork?

- Is the existing process owned and managed by a part of the organisation heavily involved in the artwork process?

- Would the owners of the current process consider artwork an appropriate extension of their scope?

- Is the existing process governed by an appropriate steering team that will take fair account of the needs of the artwork process when considering changes to their process?

- Is the artwork capability sufficiently small in scale to be successfully managed within another support process?

If the answer to any of the above questions is no, then careful consideration should be given to creating an artwork-specific support process rather than trying to force fit artwork into an existing process capability.

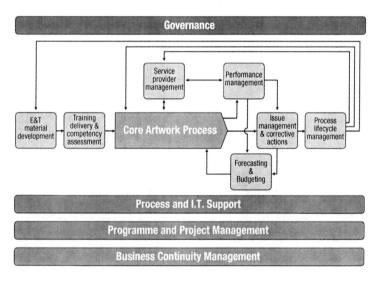

Figure 6.1 Supporting processes model

We will now discuss each of these supporting processes in a little more detail. What follows is not intended to be an exhaustive analysis of each of the processes; the discussion is aimed at making you more aware of each supporting process and, in particular, some of the specific nuances associated with artwork capability.

Governance

Perhaps the most important supporting activity is that of ensuring that the holistic artwork capability is governed effectively. Indeed, we would argue that unless an organisation gets the governance right, the core artwork process will not be sustainable in the short to medium term.

Let's start by looking at the purpose of governance. Figure 6.2 outlines this for an artwork capability. A particular organisation's view of governance may vary, but we are sure you will recognise the key elements of governance in this model.

Key to getting the governance of the artwork capability right is to recognise that the process is, at its core, a cross-functional, international and in many cases cross-organisation process. Unless all key functions, geographies and organisations are represented in the governance of the capability, it is highly unlikely that effective decisions about the capability will be made, or that there will be effective buy-in from across the impacted parties to ensure the effective sustainable operation of the capability.

Therefore, we would recommend forming a cross-functional steering group to govern the artwork capability on an ongoing basis. It is our experience that this is often a new steering group, as existing groups either do not have the appropriate cross-functional or geographical representation, or do not have the bandwidth to do the job effectively. The members of the steering team should be selected at an appropriate level in the organisation to have the authority to carry out their responsibilities on the steering team and represent their function/geography effectively. Furthermore, we would recommend avoiding populating the governance bodies with individuals who are also members of the teams implementing initiatives associated with the artwork capability; this can often lead to an unhelpful conflict of interest and consequent reduction in ambition.

Example Governance Purpose
Ensure a clear vision is defined and communicated
Approve artwork capability design
– Scope
– Principles
– Process Design
– Organisation Roles & Responsibilities
– IT Capabilities
– Facilities
Ensure resources are in place to deliver the current capability
Ensure process performance is acceptable
Ensure appropriate approval is sought for initiatives
Ensure an initiatives prioritisation framework is in place
Ensure resources are in place to deliver the initiatives
Resolve stakeholder conflicts

Figure 6.2 Example governance purpose

Having established a governance team, the starting point for effective governance is to ensure that there is a clear vision of what the artwork capability needs to be now and its development path in the future. It is important that this vision has buy-in from the whole steering team and is effectively communicated to all impacted stakeholders in an appropriate manner. This communication step will help to ensure that the whole organisation is aligned behind the development of the artwork capability on an ongoing basis.

Once the vision is in place, the steering team need to ensure that appropriate initiatives, or projects as many organisations term them, are developed and approved to achieve the vision in an effective and timely manner. Once approved, steering team members need to ensure that appropriate resources from their parts of the organisation are available to support their delivery. The steering team is responsible for governing the individual initiatives as well as the overall programme of initiatives to ensure the delivery of the overall vision.

Having created an artwork capability through one or more initiatives, the steering team then needs to govern the ongoing operation and development of the capability over time.

Process lifecycle management

We have described at length the core artwork process. In order to develop and sustain this process, there needs to be a formal process to manage its ongoing lifecycle. This process needs to manage the definition, deployment and maintenance of the artwork process and, potentially, certain key supporting processes. Figure 6.3 outlines the key steps in this process.

Figure 6.3 Process lifecycle management process steps

Continuing a key theme of the governance discussion, the fact that the artwork process is a cross-functional, cross-geographical and often cross-organisation process needs to be reflected in the way in which process design changes are assessed, approved and implemented. In the same way that we suggest that the steering team needs to have cross-function and cross-geographical involvement to ensure a robust and bought-into solution, so should the working team that develop any detailed proposals and approve the detail of any changes.

Given that the artwork process interfaces with a significant number of other processes in a typical organisation, it is necessary to ensure that an effective impact assessment of any proposed change to the process is made to ensure there are no adverse impacts on any interfacing processes and that any necessary corresponding changes to the interface points with other processes are managed effectively.

During the detailed design of the change, a risk assessment may also be necessary to ensure that the proposed changes are robust.

Authorising the change to the process needs to involve representation from all involved stakeholders in order to ensure a robust solution and buy-in to the change. This may not involve signatures from all impacted parties, but it should ensure that all impacted parties have been effectively consulted during the development and approval process. This is a very easy thing to miss in the rush to implement an urgent improvement and will invariably adversely impact the sustainability of any proposed change.

Having approved a design change, the formal process documentation and, if appropriate, associated education, training and competency assessment material needs to be updated and approved.

Finally, implementation of the change needs to be managed effectively. Two issues need to be considered in particular at this point. Firstly, is there a need to test the change in any way before putting it on general release? The answer to this question will depend on the result of the impact assessment of the change and the risk the change introduces. In many cases there will be no requirement for testing of the change, but in some cases there will and for these there needs to be a managed process in place. Secondly, consideration needs to be given at this time to the impact of the change on work in progress. With the majority of simple process changes, it will be possible to adopt any changes for all changes that are currently under way immediately. For more complex changes this may not be possible and special treatment will need to be given to work in progress during the implementation of the change. It is at these times of transition that the artwork process is most vulnerable to errors, and therefore these transition periods should be carefully and formally managed.

Education and training material development

Having defined a process and created the appropriate process documentation, it is time to prepare the education and training material that goes with it to ensure the process operators are fully competent in their roles. Clearly, the process documentation itself will normally form a core element of the training material, but it needs to be augmented with context, examples etc. Figure 6.4 outlines the key steps in this process.

Figure 6.4 Education and training material development and maintenance

The start of the process is to identify the training needs of each role in the process. This involves analysing the tasks that each individual role needs to perform, then identifying the skills and knowledge required to perform them. This will then lead to an understanding of the education and training that will be required, along with any competency testing that needs to be performed.

Having understood the education, training and competency testing requirements for each role, it is time to define the structure of training modules. When deciding how to structure training modules, a number of factors should be taken into account:

- All individuals need to understand the overall end-to-end process in order to understand their role in it and have appropriate empathy for the other process participants. Therefore, there should be some form of overview process training that all users receive.

- In our experience, the most effective task training is that which focuses on a particular individual's role in the process. Therefore, we would recommend that any training programme should be designed to ensure that, as far as is possible, an individual gets the training they need to do their job and do not receive training for tasks they will not perform. If individuals have to sit through a lot of training which is irrelevant to them they will be frustrated from day one. The temptation of the team developing training is to make life easy for them by building a small number of training modules that everyone has to go through in one form or another. We would suggest that this approach is rarely in the best interests of the organisation as a whole.

- Training material should try to be captivating for the audience. The training should include real world examples, exercises and external context-setting material wherever possible. Again, there is a great temptation on the part of the people developing the training to cut corners here and create training which immediately turns users off the process, before they have even begun to use it.

- Training material often needs to be built with three audiences in mind. Firstly, there is the initial group of people who will be involved with the launch of a new process or significant change. Secondly, there are those individuals who join the process sometime later, often one at a time. Thirdly, there are all the people we have just talked about who will need either periodic refresher training, or will simply want to refer back to the training when they are executing the process. For these reasons, the format of the training materials and the mechanism of delivery should be considered hand-in-hand. Classroom training by a trainer to a large group may be efficient on day one for most users, but it is rarely the basis for a sustainable training model.

- Consider the mechanisms that are already available to users to receive training. Tools such as corporate intranet-based training delivery services can be very effective, as can things like recorded training sessions or recorded commentaries over slide presentations. A point to consider here is that any recorded material can be offered to users to be consumed at a later date when they need it, without tying up valuable trainer resource, or worse still not being able to deliver critical training due to a lack of trainer resources.

- Do not forget that there are likely to be a number of users involved in the process who will not be employees of your organisation. Therefore there needs to be a mechanism to identify their training needs effectively and deliver appropriate training and competency assessment of these individuals.

- There will be some core skills training required by one or more roles. Topics such as basic project management and effective team-working come to mind as obvious topics for a number of roles involved in the artwork process.

- Where education and training material is being assessed for a change to an existing process, two things need to be considered: difference training for existing users and the update of the core training material for new users.

- For large organisations, with many hundreds if not thousands of users involved in the overall process, automated tools to drive the creation of individuals' training curriculums, step them through the training process and manage competency assessment can be a significant benefit and should be considered.

- There needs to be an audit record to demonstrate compliance to auditors. This needs to include items such as an individual's training record and evidence that competence has been achieved before the individual is allowed to carry out tasks in the process.

Having taken into account all of the above drivers and probably some others pertinent to your organisation, a training module and competency assessment structure can be developed. Then it is a matter of creating and approving the training material.

Finally, as has been touched on above, it must not be forgotten that much of the training material needs to be maintained as the process changes over time. Many organisations use the technique of defining expiry dates on procedures and training material to ensure that it is reviewed and updated regularly. This ensures a sustainable process in the medium to long term as process operators come and go through the natural course of business life.

Training delivery and competency assessment

Having developed the new or revised business process documentation and associated training material, this needs to be delivered though a managed process. This is particularly important in the GxP world of artwork, as users will not be able to perform their part in the new process until they have completed their training and demonstrated their competence.

If we consider the implementation of a new or significant change to an existing process in a large company, the activity of managing the training delivery and competency assessment will be significant, probably requiring a dedicated team and supporting tools. Identifying and managing hundreds of users from all over the world through significant amounts of training materials and then managing the competency assessment of them is a significant project in its own right and should be treated as such.

At minimum, this activity needs to identify the impacted user base by individual name and actively plan and track progress of education, training and competency assessment. For smaller organisations this can be done by a small group using simple spreadsheet-type tools. For larger companies, the existing training departments may need to be actively involved and specialist tools may need to be deployed.

Another consideration that we will introduce here is that of managing key resources during significant change initiatives. All change projects seem to end up relying on a small number of individuals with deep subject-matter expertise. They are the experts to whom, for good reason, everyone needs to go to discuss any change and get their input. They are key to developing the new processes, instrumental in developing the new process documentation and training and now are called upon to deliver much of the training. If this situation is not actively managed, they will soon become the bottleneck of the change activity. Furthermore, as project timelines often slip and unanticipated issues arise, these critical resources are left having to spend more and more of their time helping the project catch up, rather than doing what they had planned to do, further exacerbating the situation. If not actively managed, something will break sooner or later.

In a smaller organisation, it may seem that the issues are not as great, but often, whilst the scale of the problem is smaller, so is the resource available to manage it, as smaller companies have fewer people to call upon.

Issue management and corrective action implementation

For any operating business process there will be issues that occur. The artwork process is dealing with an ever-changing external legislative environment and therefore is constantly having to deal with new things. Furthermore, the business environment in which most artwork processes are operating is changing, companies are acquiring new assets, consolidating, changing organisational structures etc and all of this can have an impact on the ability of the artwork process to operate. In an ideal world, many of the issues would be planned for ahead of time and changes introduced to cope seamlessly with changes before they become issues. Back in the real world however, these changes often simply manifest themselves as process issues and/or failures and they need to be managed in a structured way. Figure 6.5 outlines the basic steps in an issue management process.

Figure 6.5 Issue management process

It may seem obvious, but the first step of the process is to put in place some mechanism to capture issues as they occur and ensure their existence is communicated effectively. This process would also normally involve some form of prioritisation to ensure that business-critical issues get managed in a timely way. It is worth also remembering that there needs to be a culture in place that encourages the identification of issues.

The second step involves an initial assessment of the issue to allow it to be assigned to the appropriate resolving person or group. Only then can someone work on the issue.

Many larger organisations will overlay on the above approach a first point of contact resolution approach. The basic principle is one of Pareto: 80% of the problems can be resolved with 20% of the (simple) solutions. Therefore, if the first point of contact for someone with an issue is able to deal immediately with 80% of the issues, then the whole mechanism to manage the resolution of issues can be significantly streamlined. Organisations will put in place help-desk operations equipped to deal with what are often termed level 1 issues, e.g. password resets. If the help-desk cannot resolve the issue, they are able to identify the most likely resolving agency and refer the issue to them.

Once with the appropriate resolving group, two activities typically occur.

Short-term issue mitigation

When issues arise, there is often a pressing need to ensure that business continues. Therefore, it is often appropriate for a short-term work-around to be developed to enable the continuation of activity.

Such short-term work-around activity should be subject to careful consideration and may even warrant risk assessment. Once developed, the work-around should be documented and communicated to the necessary users. For a more significant work-around, it may be necessary to plan and manage the deployment of the work-around.

Some organisations put in place a process-alerting mechanism to make relevant users aware of potential issues they may face and advise them of temporary work-around procedures.

Once in place, the issue and its work-around should be monitored. Consideration can then be made as to if and when a longer-term resolution to the issue is appropriate.

Long-term issue resolution

It is always preferable to eliminate issues in the medium to long term wherever practical. It is important before making any change to the process to understand the root cause of the issue. Without this knowledge, it is highly unlikely that the appropriate corrective action will be taken. Having established the root cause of an issue, the process lifecycle management process can be triggered to design and implement the necessary change.

A final point worth noting is that issues come from users, who need to have their issues resolved in an effective and timely manner. The direct consequence of not managing this resolution effectively is that business will not get done on time. The indirect and potentially more serious medium- to long-term issue with not managing this situation effectively is that users soon get frustrated by the process and potentially work around it.

Performance management

Once established, the performance of a new artwork capability needs to be actively managed to ensure that it is meeting the ongoing, and as we have already stated, often evolving business needs. The old adage "not measured, not managed" applies well to the artwork process. The process has many steps, with many different individuals involved and often several iterations. Without some good key performance indicator (KPI) reporting, it is next to impossible to understand what is really going on and therefore how to make improvements.

If you are making a significant change to an existing process, it is also worth collecting some basic performance measures of the existing activity before the change is made. This will act as a "baseline" on which improvements in the new capability can be measured. Having this baseline information can go a long way to demonstrate the effectiveness of new capabilities.

The object of KPI measurement should be to give the management team the information that tells them what is happening and allows them to identify the root causes. We suggest the following areas should be considered when defining KPIs for an artwork process.

- Right first time for the overall process.
- Right first time for key steps of the process.
- End-to-end process execution time.
- Step process execution time.
- Schedule adherence.

- Process failure reasons at key steps.
- Cost.
- Waste.

Another aspect of KPI data collection to consider is who is doing the activities. Collecting this information can be particularly useful in identifying individuals or specific locations which are having difficulty with the process, giving the process management the opportunity to take the appropriate corrective action.

Be careful not to create a small industry out of the process of measuring KPIs. We have worked with a number of clients for whom the exercise of collecting KPIs appeared to be the end result, partly because it was such a difficult task for them in itself. KPIs are a management tool to be used in helping decide what corrective action needs to be taken; pragmatism is needed to balance the effort involved in data collection with the benefit derived.

Figure 6.6 outlines a process framework that could be adopted to manage formally the performance of the artwork capability.

Figure 6.6 Performance management process

The starting point of the process is to define the performance targets for the next period; in most businesses this will be for the next year. Clearly, having established your KPIs, these targets will be largely KPI-based targets. We would suggest that taking a balanced approach to performance target setting is sensible. The targets should cover:

- Quality performance.
- Business performance requirements.
- People performance and satisfaction.
- Financial performance.

Another word of caution is appropriate here: the artwork process is highly people-centric and people have a tendency, often usefully, to ensure that they deliver their targets. Before setting targets, it is worthwhile thinking through the potential secondary effects of particular targets to ensure that, if followed to the letter, a particular target does not create serious undesirable effects. One way of mitigating this risk is to ensure that when communicating the KPIs and targets, the potential undesirable effects are explained.

It should also be noted that, given the highly people-centric nature of the artwork process, there is ample opportunity for individuals and even groups to manipulate their KPIs. One way to avoid this is to ensure that the KPIs, and particularly the way in which they will be measured, is tightly defined up front. This gives people less opportunity to interpret measures to their own benefit.

Having defined KPIs and targets, the management team are then left to gather the KPI information periodically, synthesise it, publish it and take decisions on what, if anything, needs to be done to correct any issues that are arising. It may also be appropriate, when faced with an issue, to put in place temporary measures which allow you to understand the issue. Once understood, the measurement can stop.

Programme management

It may seem somewhat strange to you that we have included programme management in the suite of supporting processes for an artwork capability. The reason we do this is relatively straightforward. Corrective action and improvement activity associated with the artwork process will fall into two broad areas: those activities which can be completed with a very small number of people in a very short period of time and those which require a longer period of time and more resources to make them happen. For the purpose of this discussion, we will call the former, just do its (JDIs) and the latter, projects.

One would expect that JDIs are able to be managed by the normal artwork capability support team without the need for special management. Ensuring there is some tracking of the outstanding JDI issues, the length of time they have been open etc, will be necessary.

The projects are a different matter. A number of organisations have had a great deal of success in considering any more significant or longer-term improvement or development activity as a discrete project. Each project has clearly defined benefits, scope, timeline and required resources. It also goes through one or more formal approval gates with the steering team. Having got a series of projects defined and some running, the steering team can then manage the programme of projects to deliver the overall business objectives. At one level, this involves steering individual projects and helping resolve any issues that may arise. At another level this entails actively prioritising projects and making hard decisions to delay or cancel some activity in the interest of delivering the best result for the business. This latter point is vital in our experience, as the artwork area seems to generate a significant number of projects which, whilst interesting to specific individuals, do not necessarily contribute significantly to the overall business objectives when compared to the other initiatives that could be undertaken.

If you are interested in understanding more about programme management, there have been many volumes published on the topic and we suggest you seek out some of these.

Service provider management

Almost all organisations will use external service providers to support some of the artwork process in some way. It is vital that these service providers be selected and managed effectively for the overall artwork process to be successful. We discuss how to achieve this in the outsourcing chapter.

Forecasting and budgeting

Having developed an artwork capability, the forecasting and budgeting process aims to do two key things:

- Ensure the overall capacity of the artwork capability is sufficient for both the forecast demand and any risks that the business wishes to mitigate.

- Ensure that enough financial resources are available to provide the required capacity and carry out any planned improvement activity.

Figure 6.7 outlines a potential process framework to achieve these goals.

Figure 6.7 Forecasting and budgeting process

Central to this process is the understanding of future demand or forecast workload and ensuring that there is enough capacity to meet it in the planning period. Some of you will immediately recognise this approach as being the same as a typical sales and operations planning process.

As with all demand forecasting, it is important that there is a focus on getting a realistic estimate of demand in a way that will mean something to those needing to use the data. Ultimately, what matters to an artwork capability is the number of artworks that the operation will have to process in the period. Therefore, getting a sales and marketing forecast in currency from the sales and marketing team is unlikely to be very insightful in and of itself. Some points to consider in this process include:

- Forecasting should aim to identify where step changing in capacity is required to allow timely management of these changes.

- Try to understand the macro changes that are occurring in the business which will have a significant impact on artwork changes. Activities such as: multi-market product launches; acquisitions and divestments; re-branding activity; supply-chain reorganisation; and introduction of new legislation are all examples of macro level business activity that can have a marked impact on the volume of artwork activity. Having understood these macro level changes and their likely timescale, it should be possible to estimate the potential impact on likely artwork volumes.

- Be sure you engage the right level of stakeholders from the appropriate functions in building a picture of the macro level business changes. If you engage with people too junior in the organisation, you may well completely miss a relatively confidential planned business change that will have a significant impact on artwork capacity requirements. Confidentiality issues may also require you to execute the demand estimation activity within a very limited and relatively senior team.

- Don't forget your right first time (RFT) performance when estimating the impact of forecast workload. If you are currently achieving a RFT performance of 50%, then for every piece of artwork that you have to process, you will need to process it twice. This will only change if you have some significant improvement activity in place which will transform this performance well within the forecast period you are considering.

- Do not be too concerned with gathering lots of granular data to support forecasts; often you will only be fooling yourself that there is more behind the estimates than there actually is. One experienced planner that we worked with was right to point out that much of forecasting is no more than informed guesswork. Bear this in mind whenever you are tempted to analyse the data in any more detail.

- Remember that the timing of macro level changes can often shift quite considerably from the early estimates of when they are to occur. You may want to run a few scenarios in your forecasting exercise to examine the implications of the potential different timings.

Having developed a realistic, appropriately high level forecast of artwork demand, you need to match this to capacity. This is where some of your KPIs will inform you. Ensure that your KPIs allow you to build a picture of the capacity and the productivity of your artwork capability and allow you to estimate the resources required to change that capacity in line with the forecast. There may also be new capabilities that are required as a result of this process which all need to be taken into account. With this information you will be in a good position to estimate the required capacity.

Once you have a picture of the capacity required to meet the forecast demand, creating budgets is a straightforward matter of multiplying the capacity plan by unit cost. Typically, organisations will have some form of budget review and approval cycle which will be used to formally approve budgets for the next financial period.

During the financial period, many organisations find it prudent to review the forecast periodically for significant changes. This information is then used to assess if there is any material impact on capacity requirements so that they can be adjusted in a timely manner.

Business continuity management

Fundamentally, this is a plan that will be enacted if any or all of the elements that make up your artwork capability are suddenly not available for some unplanned reason. Typical examples of this are building fires, key suppliers going out of business, or country/region transport issues such as happened in Europe in the winter of 2010/11. Figure 6.8 outlines a possible process framework.

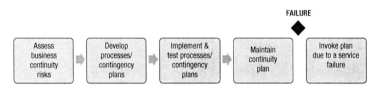

Figure 6.8 Business continuity management process

The process starts with a design phase, where the business continuity plan is developed and tested. Once in place, the plan remains inactive until one of the events that it was designed to manage happens, at which point it is initiated and should successfully manage your organisation through the event.

Fundamental to the plan is developing an understanding of the risks which may make individual elements of the artwork capability become unavailable. For each risk scenario, a mitigation plan needs to be developed and documented.

Having understood how each risk will be mitigated, all this can then be written up into one business continuity plan. There are bound to be large overlaps in the ways different risks are mitigated and it is for this reason that many organisations simplify things into one overall plan. A measure of a good business continuity plan can be considered to be: can it be used like a play book by the team who will be managing an incident, in such a way as there is little room for them to make critical mistakes or miss out critical aspect of managing the incident?

Working on a business continuity risk assessment for the first time will quite often highlight that certain significant risks are not effectively mitigated. For example, you may find that it is unacceptable to be reliant on a single organisation to create all of your artwork. Alternatively you may find that, given you have several artwork providers, it is not possible to move work between them in an emergency for some technical reason. All of these unmitigated risks should then be considered as improvement activity to be addressed over a suitable period of time.

Once created, many organisations will run mock incident tests periodically to ensure they are ready for a real incident. This may be something that is done once every few years.

Another point worth noting is that the business continuity plan needs to be reassessed whenever there is a significant change to the artwork capability. To this end, many companies add the need to assess the impact on the business continuity plan to the responsibilities of project managers who are managing the implementation of change projects.

7
Organisation

In this chapter we will discuss some of the issues that must be addressed in designing an organisation to deliver an artwork capability in a typical pharma company.

We will divide the conversation into four areas:

- Roles that support the process.
- Organisation design.
- Individual competence.
- Governance and leadership.

Roles that support the process

Roles should be structured to support the business process. Therefore you need to have defined your business process before your roles and ultimately people's jobs.

An individual role should be constructed by examining the tasks a process needs to have performed and the skills and knowledge that those tasks require to perform them successfully. Once all this is understood, a logical grouping can be performed to gather together tasks that require similar skills and knowledge.

Figure 7.1 lists the typical roles that result from this analysis and are required to support the core artwork business process we described in Chapter 4.

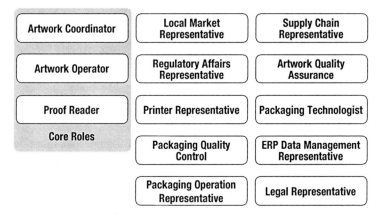

Figure 7.1 Core roles

When discussing roles and people's jobs there is often confusion that arises. To clarify, a role that supports the process does not necessarily equate to a particular individual's full-time job. Indeed, in many organisations, most of the people involved in artwork processes will have several roles in other processes as well. Let's take an example to illustrate the point. Consider a person with the job title of regulatory labelling lead. This person's main job is to define and negotiate regulatory labelling text with external regulators and therefore has one or more roles in the processes associated with generating agreeing this text. A small part of their activity is actually directly involved with the artwork process in providing the text documents and reviewing artwork results. Hence you can immediately see that their "role" in the artwork process does not equate to their "job".

In a typical organisation there are a number of artwork roles that tend to equate to full-time roles for individuals, or at least absorb a very significant portion of an individual's time. This is a point that will become useful when we discuss organisation design later in this chapter. These "full-time" roles are typically:

- Artwork co-ordinator.
- Artwork operator.
- Proofreader.

Organisation structure

Having looked at the key roles that support the process, we can now start to look at the organisation structures that will best support the new capability. We will focus here on the roles and people who spend the majority of their time carrying out the artwork process. For those people who only spend a small amount of their time carrying out artwork process related tasks, it is normal for them to remain within the structure of their current organisation and we will therefore not consider them any further here.

To illustrate some of the issues and organisation design considerations, we will describe the situation that can be found, to a greater or lesser extent, in large pharmaceutical companies at some point in their development.

A typical large pharmaceutical company sells products in over a hundred countries in the world and has a small number of regional management operations centres. Its supply-chain uses tens of factories, some internal and some third party, to package products. Often, most of the packaging plants will be product- rather than market-focussed due to a mix of technical constraints and organisational history.

Historically, the artwork creation capability of these pharmaceutical companies has been driven by the operations that need packaging on a daily basis – the packaging factories. If artwork is not created correctly and on time, the packaging factories are often the ones where the issues manifest themselves. Understandably, therefore, they traditionally took control of the situation and put in place their own artwork capabilities. This understandable logic led to an artwork studio being located at most of the packaging factories, and specific relationships being put in place for any sites where it was not appropriate. The resulting situation required many artwork studios to co-ordinate with most of the markets, regulatory teams, legal teams and, in many cases, a subset of the same packaging suppliers.

From the perspective of the packaging factory management team, this model can appear effective because they have a great deal of control. If there are issues with the supply of packaging components, their team can normally resolve them, and if this is not possible it was always clear which external department needs to take corrective action.

For the many people in the markets and other functions who are involved in the process, this situation looks very different. For a start, they have to deal with many artwork studios, each effectively operating their own process. Because each of the studios has a different artwork process, they have their own training approach, materials and training support team. This means that for each artwork studio they deal with, they have to go through a new set of training and competence assessment. Furthermore, the different studios often have no common way of documenting the process or going about the training and competence assessment.

To make matters worse, the people in the markets and other functions often only deal with artwork occasionally. This can mean that when they come to work on a piece of artwork, they are dealing with an artwork studio with whom they have not worked for a very long time. They find themselves having to develop new relationships with the people in the artwork studio and remind themselves of the specific process that the particular artwork studio uses. Often, artwork is needed in a hurry and there is not enough time for all of this preparation work, so individuals do their best and carry out their tasks as they remember them from the last time they did an artwork. Typically, this leads to a number of recalls over the years which result from people outside the artwork studios not doing what they were supposed to do in the individual artwork studio process.

This situation is not ideal in the factory based artwork studios either. To get the process working effectively, each party needs to try to develop a working relationship with the market, regulatory group and supplier involved in the process. They therefore spend a lot of time and energy developing and maintaining these relationships. Typically, they end up focusing their energy on the people who spend most of the time doing artwork-related activity, the large market-related people and the large suppliers, and these people often work well with their process. However, they often have problems with the smaller markets, which never seem to be able to do what is expected of them correctly or on time.

Looking at the artwork studios as a group, it was clear that there is a significant amount of repeated and potentially non value-added work going on. Effectively, each studio is duplicating the following activities:

- Developing and maintaining relationships with individuals in the markets, other external functions and third party organisations.

- Training and ensuring the competence of these external individuals in their processes.

- Correcting issues because the external individuals did not follow their process correctly first time.

- Understanding the subtlety of individual country artwork-related packaging legislation requirements.

Furthermore, given that each studio operates independently and there is no common IT platform, they often have no meaningful knowledge of the packaging designs done by other factories for the same markets. This leads to inconsistent artwork brand images existing in each market.

The senior management of the organisation also find the situation challenging. On a day-to-day basis, there is no consistent picture of what is happening with artwork changes across the business. If changes need to happen in several different packaging factories, the situation is particularly difficult. Often, business change management teams have to put in place their own project co-ordination teams to understand and co-ordinate what is going on and ensure they get the results they need.

Furthermore, whenever issues occur in the artwork process, or changes need to be made across the company, it is very difficult, if not impossible, to ensure that the appropriate corrective action has been taken in all the artwork studio processes. Regulatory requirements, such as being able to report on the status of critical safety changes, becomes very time consuming as the information is not readily available in one place and every artwork process is different.

The situation we have just described, whilst it has its merits, is very inefficient for the company overall. At worst, it presents many opportunities for errors and omissions in artworks that will eventually lead to recalls.

Clearly, if you have a very small or highly market-aligned supply-chain then the issues and risks may not be as great, but many of the same issues will present themselves, just on a smaller scale. For small organisations, any inefficiency is a serious issue as they are often very pressed for precious resource.

So how can a company organise itself differently to resolve many of these issues?

One topic that we have already discussed is that of creating a single core process, which can be operated in the same way at any node. This will clearly ease some of the issues that we have just discussed. We would recommend this is backed up by a appointing a single end-to-end process owner role to help enforce and maintain one way of working across the organisation.

From the roles discussion earlier, you will remember that there are three roles in the process we present here which are typically staffed by individuals who spend the majority of their time focussed on artwork activity:

- The artwork co-ordinator.
- The artwork operator.
- The proofreader.

Co-locating these roles and putting them under the same management helps to drive some level of common ways of working through the organisation. Indeed, if you look at the way the site-based artwork studios have usually evolved, these are exactly the roles that exist there, together with the management structure to support them.

Creating one or more artwork studios to serve the whole organisation can have a number of key benefits which have proved to be very powerful in some organisations:

- With one or a small number of artwork studios driving the global artwork activity, it is much easier to create and maintain a truly single global process.
- The need to develop far fewer relationships results in higher quality relationships being formed, which in turn results in the process working much more effectively.
- Individuals in the consolidated operation can leverage their specific knowledge across many more artwork changes.

- It is much easier and more efficient to provide all users with the training and support they need to carry out their activities correctly.
- Furthermore, with a critical mass of key roles at the artwork studio, higher quality training and competence development becomes possible.
- Awareness of the overall picture is improved, allowing improvements in things like brand consistency.
- A larger artwork studio leads to benefits of scale in support services and management overhead.
- Improvement and change activity is much easier to implement as there are fewer nodes to deal with. Furthermore, the few artwork studios that do exist can operate as powerful drivers of the implementation of change.
- Gathering and reporting consistent status and performance information is much more achievable.

When deciding on how many service centres are needed, a number of factors need to be taken into account, all of which will be very different depending on the situation within each company. The types of things which impact the decision include:

- Number of countries in which products are sold.
- Commercial, supply-chain and support functions organisation structure.
- Existing resource levels and the potential impact of reorganisation.
- The political will for change.
- The budget available for reorganisation.
- The quality and extent of the IT tools available to the artwork process.

Individual competence

Ultimately, the quality of the output of the artwork process depends on the competence and diligence of the people executing the process. Competence depends on a number of factors:

- The individuals' innate ability and temperament.
- The individuals' relevant experience.
- The quality of the training they receive.
- The quality of the tools they have at their disposal.
- Ensuring they are appropriately competency assessed before being allowed to carry out their role.
- The culture they are working in.
- The support they get to resolve issues that are bound to arise.

Selection of individuals suitable for a role in question is always a key first step for any role or job. It is therefore important to define the skills and capabilities that an individual must possess in order to be able to carry out a particular role successfully.

It is tempting, particularly when considering the staffing of the artwork studios, to think that any of the individuals available can perform any of the roles. In our experience, this is rarely the case. Artwork operators, proofreaders and artwork change co-ordinators tend to have very different skill sets and aptitudes. For example, proofreading requires a very process-centric, detailed, focussed skill set and a person who is very happy concentrating and working alone for many hours at a time. In contrast, artwork co-ordinators need to have a high degree of organisation and co-ordination skills as they are effectively project managers for each artwork change. This needs to be backed up by a strong ability to form relationships and work in remote teams, whilst managing issues as they arise. People in either of these roles are unlikely to have the aptitude or interest in performing each other's role.

We discuss training in other chapters of this book, but we will re-emphasise here that the quality of the training that individuals receive is key to their competence in an area such as artwork creation, particularly as the process is so susceptible to risks from apparently small errors. For many organisations, the training associated with business processes is limited to reading through the standard operating procedures and signing that they have been understood. Perhaps there will be a questionnaire to check understanding, although this often simply only serves to verify that the whole of the procedure has been read.

Training for the key roles in the artwork process needs to go much further than this, involving a combination of techniques to ensure sustainable competence of the staff. Below we list some of the steps in this development of sustainable competence that we have seen used successfully. These include:

- Developing a clear sense of the importance of correct artwork to patient safety and supporting this with a culture of patient safety being the first priority.

- Explaining or demonstrating some of the inherent human limitations which make a particular task more difficult.

- Demonstrating the typical errors that occur.

- Ensuring that staff have a clear understanding of the business process and the detail of the tasks they are required to perform.

- Providing education and training materials and support in a number of different ways that suit the individuals and their involvement in the process. For individuals who only perform artwork activity infrequently, education and training which is available when they need to do the tasks and is formatted to help them through the work can be very useful.

- Mixing delivery techniques to suit the audience and the skills being developed.
- Shadowing and one-to-one coaching over extended periods of time for complex roles or tasks.
- Providing job aids such as checklists to ensure that critical aspects of tasks are performed every time.
- Recognising the language limitations that some people face and adjusting the materials and delivery techniques accordingly.
- Recognising the different cultures of individuals and teams involved in the process and adjusting the education, training and competence assessment approach accordingly.
- Developing a culture which focuses on right first time and welcomes mistakes, however minor, as opportunities for improvement and not as things to be hidden or covered up.

When we talk about training in this context, we are referring not only to initial training, but also refresher and update training and ongoing competence assessment. Without this ongoing activity, the competence of staff and the effectiveness of the overall process will decay.

We have discussed job aids elsewhere, such as checklists, or perhaps IT tools which ensure key steps are performed or that data is entered with a certain format. All these are key aids to improve the resulting competence of the operators and, ultimately, the effectiveness of the process.

Competency assessment is perhaps one of the areas that is most difficult to do well. An approach which looks at the criticality of the role and the complexity of the tasks will help identify areas for special focus. Another area for attention is those people who only perform tasks infrequently. Competency of these individuals should be critically assessed both initially and on an ongoing basis. One word of warning here: it is also important to be realistic about the competence that can be achieved by an individual who only performs tasks infrequently. Recognition of this limitation may even lead you to restructure the way work is done. There are various ways to assess competency, a topic which is beyond the scope of this book.

Leadership and governance

Given the cross-functional and cross-organisational nature of the artwork capability, establishing the right inclusive leadership and governance is key to the long-term success of the capability. All stakeholder groups involved in the delivery of the artwork capability need to contribute effectively or the whole process will fail. Therefore, all parties must buy into their role in the process and actively contribute to it. This will rarely happen if they are simply passive bystanders in the design of the capability or the delivery of the resulting activities.

Role of an Artwork Governance Team

- Ensure that a clear vision and strategy is defined and communicated
- Approve the artwork capability design
- Ensure that the performance of the artwork service is meeting business needs
- Ensure improvement activities are prioritised and approved
- Ensure resources are in place for the artwork service and improvement activity
- Ensure stakeholder group conflicts are effectively resolved

Figure 7.2 Role of an artwork governance team

As we discuss elsewhere in this book, we would recommend establishing a cross-function governance team to steer the establishment, ongoing delivery and development of the overall artwork capability. This governance body should include membership from all stakeholder groups involved in the process, including, where appropriate, external service providers. Figure 7.2 provides some thinking on the role of such a governance group.

It is all too easy when forming and managing governance teams to focus on the steering and decision-making aspect of the activity. If you are not careful, this may result in the leadership responsibilities of the team being overlooked. The governance team needs to ensure that they provide leadership to the artwork function in a number of distinct ways. Firstly, they need to ensure that a vision and strategy for the artwork capability is developed, agreed across all impacted stakeholders and communicated effectively to the broader organisation. Secondly, they need to ensure that the journey to achieve this vision is structured and managed effectively and that progress is communicated to the wider organisation. Thirdly, the leadership of the governance team needs to manifest itself in decisive decision-making that supports the vision and goals of the artwork capability. Finally, the behaviours the leadership display need to actively model and support the key cultures that underpin the successful service delivery.

To support these leadership activities, some organisations purposefully put in place a number of key roles:

- Senior sponsor – a senior member of staff who will represent and support the overall artwork capability at the highest levels in the organisation.

- Governance team chairperson – the leader of the governance team who ensures that the governance team activities are managed effectively.

- Artwork process owner – an individual who is responsible on a day-to-day basis for ensuring that the end-to-end artwork process operates effectively and that any improvements to the process are appropriately designed.

With all of this in place on an ongoing basis, the artwork capability should remain effective and appropriate for an organisation over time.

8
Technology

In order to examine the areas where technology can help an artwork capability, we will start with a discussion of some of the key issues that arise when trying to manage an artwork capability in a low tech, ostensibly paper-based world. We will then go on to describe the principal areas of artwork technology capability one by one.

Limits of the low tech world

Here we will discuss some of the principal issues which are faced when carrying out artwork activity that can be assisted by technology.

Basic human error
People, by their nature, make mistakes. No matter how skilled we are, no matter how well trained, no matter how diligent we try to be, people make mistakes through distraction, stress, boredom and many other factors. This fact is well known and is one of the reasons why GMP processes typically require two people to verify critical activities. Artwork involves a lot of detailed and repetitive tasks; therefore, verification of activity or automation of tasks by technology offers opportunities to reduce these mistakes.

Transcription errors

A specific cause of error in the artwork creation process is that of transcription errors. A lot of the artwork process involves taking information from one source and putting it, or transcribing it, into another document. This relatively mundane task of copying information is an area prone to human error. Therefore, technology can provide help in reducing the possibility of transcribing errors through simple tools such as copy and paste.

Proofreading errors

We discuss the significant issues associated with proofreading in the core process chapter. Technology can offer a number of tools which:

- Reduce the number of missed errors.
- Reduce the time to do the proofreading activity.

Information transfer lead time limitations

The artwork process inevitably means that people from many different functions and organisations, in many different locations and countries, need to share the same information and communicate effectively in a timely manner. In a "low tech" world, the logistics of providing the right versions of documents, in a quality that is acceptable for the activity being performed, is very challenging. The time involved in passing physical documents between locations increases the overall lead time of the process. The challenges that this presents are often so fundamental as to shape the organisation structures and processes themselves to mitigate otherwise unacceptable process lead times. Technology allows us the possibility to pass information reliably between people instantly, improving process lead times and freeing organisation design from such information transfer constraints.

Information availability issues

We often find that if information and documents are not readily accessible to people, they will naturally tend to get around the issue by collecting their own store of information, or by using alternative uncontrolled sources of information. Clearly, in a GMP process, this quickly leads to a situation where the quality of the critical information in documents and the validity of decision-making can be put at risk. Technology provides an opportunity to provide single sources of information to all those who need it, no matter their location or the organisation they are in. Clearly, along with access to information comes the issue of segregation of information and security, which must also be addressed.

Process adherence issues

It is important to ensure that all the relevant steps of the process are being carried out in the right order, so that a quality result is achieved. In a manual world, this often involves a quality control review of the documentation that results from the process to ensure there is evidence that all the critical steps were carried out and the sequence of the critical activities is correct. If errors are found, then corrective action needs to be managed, often resulting in significant recycling of the process, which in turn impacts lead times and efficiency. Workflow technology offers the opportunity to enforce process step sequences, therefore eliminating the possibility of missed steps and the checking required to verify this has happened.

Co-ordination difficulty

Furthermore, because many people are involved in many different steps of the process, significant co-ordination effort is normally required to ensure things happen on plan. Many organisations will establish one or more roles which are employed to do nothing more than progress artwork changes, ensuring the right people are doing the right activities in the right sequence. The same workflow technology significantly reduces the need for co-ordination and routing activity, freeing staff up to spend more of their time dealing with the many real issues that are bound to crop up.

Personal work management issues

For individuals involved in the process, knowing when they need to do things, ideally ahead of time, will lead to a more efficient and shorter end-to-end cycle time. Furthermore, it will allow them to plan their work, allow quality time for quality critical work, feel more in control of their environment and often lead to a greater sense of satisfaction. Technology provides the opportunity for people to have visibility of the work coming, or due now, and alert them of overdue activities. Knowing when tasks are going overdue is also knowledge which is valuable to the management of the process as, armed with this information in a timely manner, specific corrective action can be taken to keep timelines on track.

Activity tracking and performance measurement issues

The artwork process involves many steps, performed by many different people. In a typical organisation, there are many artwork changes occurring at the same time. It does not take a very significant amount of activity before it can become very difficult to keep track of activities, know where things are at present and where to focus attention next. Collecting information for performance management KPIs also becomes very challenging. Without technology, this sort of tracking activity needs to be done on paper, perhaps using a visual office approach on the wall. One drawback of this approach is typically that the progress management is only done at a macro level, as tracking at the individual task level becomes unmanageable. Also, the tracking of information is rarely real time, so does not provide timely information for corrective action to be taken. Lastly, the tracking information is not readily available to all people impacted, so its potential value to the organisation is reduced. Technology offers the opportunity to have this information available to everyone in real time.

Lack of information visibility opportunity cost
When information and documents are locked away in filing cabinets, or in individuals' own ad hoc documents or file stores, they are not readily available to others. There are many unplanned benefits that stem from making information available so that it can be accessed when people want to see it, not just when a particular process dictates that it is provided to them. The internet is the largest example of this and has clearly resulted in many unforeseen benefits.

Summarising, we can quickly see that technology presents a global, cross-functional, cross-organisation process such as the artwork process with a number of significant opportunities:

- Improved quality of key activities.
- Improved process adherence.
- Improved audit trail.
- Reduced lead time.
- Single source for information readily accessible to all.
- Reduced low-value-added activity and waste.
- Greater work satisfaction.
- Greater efficiency, capacity and potentially reduced cost.
- Significantly improved identification and management of issues.
- Higher quality performance management information allowing faster, more targeted improvement of the artwork capability.

Artwork capability technologies

Having discussed some of the issues that would derive from a world with very little or no technology, we are in a better position to understand the potential application of technology that is available to support an artwork capability. We will discuss each of these in turn in the following pages.

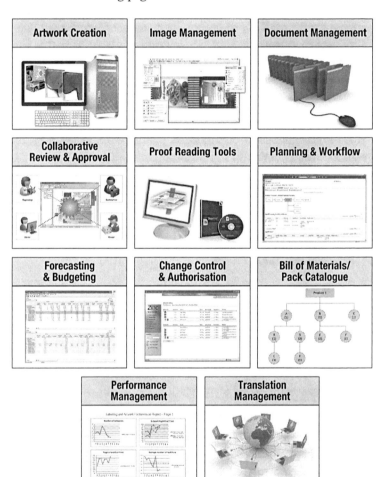

Figure 8.1 Artwork technology areas

For the purposes of this discussion, and to avoid covering the very obvious to most readers, we will assume that your organisation has the normal basic suite of business technology available to it. We will not discuss these basic capabilities any further here:

- Computers available to all staff.
- Document creation suite such as Microsoft Office.
- Corporate email.
- Server based file storage.
- Some form of intranet and internet access.

We have grouped the technology tools available to the artwork process at the time of writing this book into a number of groups, summarised in Figure 8.1. We will review each group in turn in the following discussion.

Artwork and drawing creation tools

We will start our discussion of the technology required to support an artwork capability in an obvious but often overlooked place, the tools required to create the core documents associated with an artwork capability: artwork (both graphically-rich documents such as carton designs and text-rich documents such as leaflets) and, to a lesser extent, technical drawings.

The fundamental tools for this activity are readily available industry standard software and hardware. At its most basic level, the following elements are required:

- A personal computer.
- A graphic-rich document creation application.
- A text-rich document application.
- A technical drawing creation application.

- A barcode creation application.
- A Braille creation application.
- Font management software.
- PDF creation software.
- 3D visualisation software.
- Colour calibration tools.[1]

From a hardware perspective, the choice used to be driven by the software platform that was selected, as the major software options were hardware platform specific. For those companies who were providing this capability in-house, this often led to tensions between the corporate IT group and the artwork function. Thankfully, at the time of writing this book, this issue has all but disappeared as the major software suites available can be used on any of the major desktop environments.

For the document creation applications there are many options, but self-evidently the industry leaders tend to dominate the space.

Barcode creation is an aspect of artwork creation that has become significantly more important in recent years as various regulatory authorities require barcodes of different formats to be included in artwork. Furthermore, barcodes are being increasingly used as a way to automatically identify specific packaging components in the manufacturing process. Software applications are available, either as plug-ins to the document creation software applications or as stand-alone software packages that allow the user full control over the specification and information content of barcodes. Once created, the graphical image of the barcode can readily be placed into the artwork in question.

---

1. *Many organisations choose to manage colour through the use of standard colour references such as Pantone codes, in which case the need for colour calibration tools can be ignored.*

Like barcodes, but even more recently, Braille has become a requirement for inclusion on packages by an increasing number of regulators. On an artwork, Braille is normally represented in a separate layer of a document as a series of dots in a matrix, each dot representing a single indentation that will be embossed, or represented in some other physical way, on the final printed packaging component.

Font management is something that needs careful consideration across the whole artwork process. In many circumstances, the software that we use on our computers is designed to manage fonts transparently for the user so that even if fonts are unavailable, the user will not be troubled as the software will substitute an alternative. Indeed, the invention of the web and technologies such as XML means that, by design, style and fonts are managed independently from the text content. If you have ever looked at some web pages on different pieces of hardware you will be aware of this. The internet browser in question will apply a style and associated fonts to the text content as it renders the page. This style can be influenced by user preferences amongst other things.

Whilst very useful in the user choice centric world of the web, it is not at all appropriate for software to be changing fonts on packaging artwork. Having defined what fonts appear on an artwork, we need to ensure that they remain that way throughout the whole process. For this reason, the integration of active font management applications into the artwork suite of software is essential. These applications enable the explicit management of fonts to occur across one or more applications. In an environment where many artwork creation desktops are used, the applications can be server-based and manage fonts across all desktops automatically. This technology ensures a number of key things:

- Only a defined set of approved fonts is used in artwork.
- If other fonts appear in an artwork, this will be explicitly flagged to the operator.

- An organisation can effectively and efficiently manage the introduction, deletion and payment of the fonts that they use.

It is worth pointing out that the selection of fonts for use in the artwork process needs to consider more than the look of the font and its coverage of the required language sets. Consideration needs to be given to the compatibility of fonts with other tools used in the process such as text comparison proofreading software. Furthermore, consideration should be given to the fonts used in source documentation, such as regulatory text. It is possible that if the fonts in the source text documents are not compatible with the artwork fonts, then errors will be introduced as text is cut and pasted from the source text document to the artwork. Errors such as symbol changes can occur which, in turn, if not caught, can lead to patient safety related incidents.

One last point of consideration that needs to be taken into account when managing fonts is that of licensing. Most fonts are licensed in their own right and these licences must be managed and paid for just as with any other piece of software. Corporate licensing arrangements are available for most fonts and the font management software will assist in staying in compliance with the requirements of the licences.

Given that artwork is created within a GxP process, and if errors are introduced into the artwork by the IT tools employed a recall could result, many organisations consider it appropriate to put the artwork desktop and software under the same quality and validation controls as they would any other production system.

Document image control

Paper has, for a long time, been an acceptable form of sharing information, capturing comments and recording approval through signatures. If you get a document to review on a piece of paper and then sign it to signify your approval of it, you would not question that you had reviewed a faithful rendition of the master document (as you were looking directly at it) and that there can be no mistake about which document you actually signed. The fax has, for some time now, been accepted as a form of technology which provides an accurate rendition of an original document for certain circumstances, such as approving a document via signature. When moving to electronic documents, technology needs to address three key issues:

- Is the image that a person is viewing a true rendition of the master document for the purposes in which it is being used?[2]

- If adding comments to a document, are they complete and attributable to the author?

- When approving a document or decisions is the signature truly from the person who appears to have signed it?

A couple of factors make the use of the original software that was used to create the documents undesirable:

- Software parameters can be adjusted by users that can potentially adversely impact how the document is presented to the user. Unless the software on all desktops that any user may use to review documents were "locked down", it would be difficult to be confident that the users were reviewing and approving the correct image of the document.

- Creation software is generally very complex software which comes with a high price tag. This means that it is far from an ideal solution to deploy on to many users' desktops.

Luckily, the portable document format, or PDF, standards have been developed to solve these problems. In concept at least, PDF can be

---

2. Images do not have to be 100% accurate representations of the master document; they simply have to be accurate for the purposes in which they are being used. For example, a black and white print of a document may be acceptable if the process defines that the user will not be verifying colour during their activity.

considered as the electronic equivalent of paper. No matter how the words and graphics were put on the paper, you can be confident that they will always remain the same when viewed on the paper.[3] In the same way, no matter what was used to create a PDF document, there are freely available tools available to view the documents in a manner that should display or print a faithful rendition of what was intended by the author.

As always with technology, there are a few significant caveats that must be borne in mind here. The first relates back to our fonts discussion. With PDF there are a couple of options with the way fonts are treated. They can be embedded in the document, in which case any software displaying the document will use these fonts and you should see exactly what was intended by the author. The other option is not to include the fonts in the document, leaving the software used to view the document to decide which fonts to use. Given the previous discussion on this topic, we would recommend the former approach is adopted with artwork documents.

The second issue relating to the use of electronic documents is the use and management of colour. A user could view an electronic document on many different screens, or print the document on many different printers. The user has multiple opportunities to adjust the brightness, hue and other colour affecting parameters during this process that will materially impact the appearance of colour in the image. It is for this reason that many artwork operations choose to manage colour using industry-standard colour reference codes, such as Pantone. This way, so long as one colour can be clearly distinguished from another and is rendered visibly on a screen or printout, the actual colour that is displayed is less important. An alternative to this approach is to use colour calibration equipment to calibrate all screens and printers that will be used to review artwork for colour correctness.

---

3. Clearly such issues as lighting, colour-fastness of inks etc can impact this.

Thirdly, with paper, unless someone is deliberately trying to deceive you, it is self-evident if a document has been changed, so long as it was created using indelible ink. With electronic documents this is not necessarily the case. However, using PDF technology it is possible, at the time of creating the PDF, to control what the user can or cannot do with the document. Figure 8.2 details what is typically possible to control.

PDF Control Parameter	Description of Function
Printing	Allowing or preventing the user from printing the document
Changing the Document	Allowing or preventing the user from changing the document
Document Assembly	Allowing or preventing the user to assemble the document: insert, rotate or delete pages and create bookmarks and thumbnail images
Content Copying	Allowing or preventing the user from copying the content of the document
Content Copying for Accessibility	Allowing or preventing other applications to read the content of the document, e.g. being able to select text or graphics for copying
Page Extraction	Allowing or preventing the user from taking pages out of the document
Commenting	Allowing or preventing users from making comments on the document
Filling in of Form Fields	Allowing or preventing users from completing form field information in the document
Signing	Allowing or preventing users from using electronic signatures to sign the document
Creation of Template Pages	Allowing or preventing JavaScript and other code to create new pages based on the current page. A feature which is used when completing forms

Table 8.2 Typical PDF document control parameters

All of the above discussion about PDF leads us to the conclusion that you need to decide what settings each of your documents should have and consider the use of a specialised application to create the PDF to ensure this happens consistently.

Another opportunity PDF has presented more recently is the ability to do away with native files altogether. In this context, native files are those file types normally created and used by the applications that are used to create documents, such as Adobe Illustrator files. It is now possible to use PDF documents directly in some of the applications which create the artwork. This has the distinct advantage that only one document is being managed during the creation, review and approval stages of document development; therefore, the risk of users introducing errors during the separate creation of the PDF document is eliminated. It would not be the first time for a user to use the wrong version of a native file when creating a PDF for use in the review or approval step of the process – clearly a risk to design out of the process if possible.

Despite the move to electronic documents, some users may still need to print the documents for various reasons. This could be because it is not practical to work with physically large documents on the screen; such documents would include large multilingual leaflets. It may be as a result of a request from an external party, such as a regulator, to have a hard copy of a particular document. It will therefore be necessary to print a faithful representation of the artwork on to paper or some other substrate. Given the GxP nature of the process and that the use of the document being discussed here is specifically for review and/or approval, it must be ensured that the printout is a faithful representation of the original. To this end, the printing route should be verified and controlled to ensure that the end result is correct. Furthermore, procedures should cover the printing of such documents and users should be competent in their use. Elements that need to be considered in the verification of the technology include:

- The effectiveness of the software rendering engine.

- The hardware and firmware of the printer in question.
- The ability of the above elements to produce a faithful rendition of the original document for all likely graphical and text features.
- The control of the above elements to prevent inadvertent change of any component.

Document management

The second topic we will discuss is, in many ways, the second step on the technology road path for many organisations: the use of electronic document management tools.

The basic elements of electronic document management technology that we will be discussing here are:

- Document repository.
- Document security.
- Document version and status management.
- Document visibility and transmission.
- Audit history.

The first thing that needs to be defined when managing documents of any sort is where they will be stored. Electronic document repositories can be as simple as shared electronic folders, or as complex as embedded document repositories in multifunctional applications such as product lifecycle management suites. Fundamental to all repositories should be:

- Secure location where documents will remain unaltered unless through authorised user action.
- Controlled access to users.
- Structured to allow the efficient and effective storage and retrieval of documents.

- Facility to search for documents.
- Facility to archive documents when no longer current, recording their audit history.

More advanced capabilities associated with a document repository might include:

- Defined document metadata management. e.g. product name, country etc.
- Forms management capability.
- Artwork visualisation tools.
- Automated document version management.
- Document status management. e.g. approved, pending, under revision, draft.
- Document archive management.
- Change audit history recording.
- All relevant documents instantly accessible to all users.
- Controlled file transfer capability to third parties.
- If cannot access repository directly.
- Document access control dependent upon user role/organisation.

Collaborative review and approval

For the purposes of this discussion, collaborative review and approval technology allows multiple users to see documents, comment on them either singly or as a team and ultimately approve them using electronic signature technology.

Document review capabilities generally refer to providing one or more users with the ability to view documents, make comments on them and then for those comments to be managed in order to guide the creation of a subsequent version of the document. Ideally, the comments need to be available to artwork operators, proofreaders and others in the form of a checklist to help ensure that all comments are processed effectively.

One important aspect of this mark-up technology, when it comes to reducing the errors associated with artwork, is to ensure that text comments and instructions can be linked to a location on the artwork in question. Without this, there is a danger that comments will be misconstrued to refer to the wrong part of the artwork, leading to unnecessary cycling in the process, delays and potential errors.

There are several software applications available to allow users to comment on documents. They vary in sophistication from tools which allow single users to comment on documents, to systems capable of managing the synchronous access of many users to review a document followed by subsequent consolidation and management of all resulting comments.

When considering potential collaboration platforms, as with other elements of the artwork capability technology platform, the ability of supply-chain partners to access the systems effectively should be considered. If all parties cannot access the collaboration platform, then comments from excluded parties will have to be managed by a work-around process which is likely to introduce further risk of errors and delay.

Once a user is happy with a document, they may be required to approve it formally. This is where electronic signature technology comes into play. As its name suggests, this technology allows a user to sign a document using an electronic signature. Amongst other things, electronic signature systems ensure that:

- The electronic signature is attributed to the user signing the document.

- No other user can use the same electronic signature.

- Electronic signatures can be verified for authenticity in some way.
- Audit trail information is captured identifying the signatory and when a signature was applied.

Rather than delve into the complex topic of electronic signatures here, we would refer the reader to the United States Food and Drug Administration's regulations on the topic, 21 CFR Part 11. This provides what has become the de facto standard requirement for such capabilities.

Proofreading tools

We will refer to the tools used to assist in the checking of artwork, printer proofs and printer packaging components as proofreading tools. When we discussed this topic earlier in this chapter, you will probably have recognised a number of distinct types of proofreading tool which we will discuss in more detail:

- Text comparison.
- Graphical comparison.
- Barcode verification.
- Braille verification.

Text comparison tools, as the name suggests, compare only the text content of one document against the text content of another. Within the artwork process they are used primarily at the artwork creation stage to compare source text documents with the artwork. Graphical comparison tools compare any graphical element of the artwork, including text. Graphical comparison tools are primarily used to verify that nothing has changed between the source and target document; therefore, they are mainly used to check that printer proofs and printed packaging components are the same as the approved artwork. They can also be used when modifying an existing artwork to ensure that areas of the artwork that should have remained unaltered have not been changed.

Text comparison tools

Text comparison tools are extremely useful for improving accuracy and speed of proofreading. However, at the time of writing this book they have a number of limitations which means they are aids to skilled proofreaders, rather than stand-alone, do-it-all proofreading tools in their own right.

It is necessary with all such tools to understand something about how they work in order to understand their restrictions. Let's take an example of the way several of these tools work to illustrate some of these important restrictions.

The text comparison engine at the heart of these tools effectively compares blocks of text in the source document with blocks of text in the document being compared. This raises the first restriction. Something has to tell the text comparison engine which blocks of text to compare. As of writing this book, this requires an operator to highlight source and target text blocks individually. Furthermore, if text is repeated on the target artwork several times, then the process has to be manually repeated for each block of the artwork text in the comparison process. Clearly, one of the failure modes of any form of proofreading is completely missing a block of text. In a manual world, an operator would normally use a pen to cross out all text that had been proofread to ensure the job is completed fully. Software tools need to provide an analogous way of reporting to the user exactly what text in the source and target artwork has been read and, more importantly, which areas have not.

The text comparison tools we will consider here work by comparing each letter in the source and target documents. In actual fact, they work by comparing the font character codes of each letter in the two documents. This reliance on font character code comparison is the reason that we mentioned the need to ensure that the fonts in source documents are compatible with the fonts used in artwork. This does not mean that the fonts have to be the same; it simply means that the fonts need to be compatible at the level of the meaning of the individual character codes. Luckily, a lot of work has been done over the years by font standards bodies on achieving this aim and the use of a type of font called Unicode fonts seems to address the majority of the potential issues. If used, Unicode fonts allow the use of different typefaces in the source and target documents, without concern that the meaning of characters will be confused. As an aside, Unicode fonts are also designed to give an excellent coverage of character sets for different languages, which is another significant advantage to artwork operations.

From this description of how the text is compared by the tools, you can immediately see that if your source text document fonts are not compatible with the fonts used in the artwork, then the software will report this as errors, and potentially lots of them, one for each character error. The worst situation for us is that a real fault in the text is masked by incompatibilities in the fonts, allowing a real error to get through the process undetected. The practical issue for the user is that the tool creates an unmanageable number of errors, meaning that in reality a manual proofread is the better option.

The text comparison tools can usually be configured to check all aspects of the text during the comparison and report on differences: font, typeface, bold, superscript etc. This is important in proofreading as such characteristics as specific emboldening of text can be required by a regulator.

Within a block of text, the text comparison engine has to make some assumptions about the way to "read" the text. This logic is usually quite straightforward in that it assumes a linear layout of the text and compares the characters in that linear sequence. Here we face another issue, associated with the way in which the native applications create the layout of the document, particularly where tables are concerned. There are incompatibilities with the way in which a document authoring tool like Microsoft Word constructs the text, compared to the way an artwork layout piece of software does it. This often leaves the operator with two options: break the text down and compare it on an individual cell line level and use the tool or simply proofread tables manually.

One other shortcoming of text comparison engines is that they cannot "read" text which is solely in the form of a graphical image with no associated fonts. As an example, this situation can often arise in illustrations. Here, a manual proofread must be performed.

Given the restrictions we have discussed, it is important to ensure that clear guidelines and checklists are defined to ensure that the text comparison tools are used appropriately and that manual proofreading is performed to cover any gaps.

Graphical comparison tools
Graphical comparison tools come in a number of different forms, but essentially all do the same job. They compare the visual image of a source document with the visual image of a target document. Think about placing one document on top of the other and looking for the differences and you will not be far away from imagining the way in which these tools work. There are at least two different types of tool available, one more sophisticated than the other.

The first type of tool presents the two images for comparison to the operator and interchanges them rapidly on the screen. This technique uses the natural ability of the human brain to spot movement easily. Any difference between the images will present itself as an apparent movement and the operator will be able to spot it. Clearly, this type of tool is relying on the operator's skill, concentration and diligence; in general, these tools do not produce any form of report.

The second type of tool uses a graphical comparison engine to report on differences it "sees" between the two images. This type of tool will normally report on all areas of discrepancy by highlighting the area on the image affected. The tools have varying levels of sophistication around how these differences can be processed by the operator. The tools generally produce reports.

Because these tools are comparing one image with another, there are often other issues that need to be dealt with. For example, the way varnish layers are represented in artwork can present problems. Some tools "see" the varnish layer as effectively obliterating the underlying artwork and this needs to be worked around in some way. Care needs to be taken in designing these work-arounds to ensure that other sources of error cannot be introduced.

The job of comparing an artwork with its previous version or with an electronic printer proof requires the comparison of an electronic file with another electronic file. For the comparison of printed artwork components, there needs to be a scanning mechanism employed to get the image of the artwork into the electronic domain. Companies who provide graphical comparison tools will normally either provide the appropriate scanners, or recommend appropriate scanners for use with their software. It should be noted that scanning for the purposes of this type of graphical comparison is not without its pitfalls. Characteristics of the printed components such as surface finish and crease lines can impact the quality of the scan, resulting in potential false errors being reported by the graphical comparison software. There are various techniques available to mitigate these issues depending on the specific circumstance.

As with the text comparison tools, these graphical comparison tools provide assistance to skilled proofreaders; they do not replace them.

Another area where this type of tool is being increasingly used is in packaging component printing operations, specifically on the end of the print presses. Here, every single printed component is scanned or photographed and compared with a reference artwork image. Any components which do not pass this inspection are rejected on the print press. This is clearly an excellent step forward in technology as it provides 100% quality inspection and should be encouraged.

Barcode verification tools

As discussed earlier, barcodes are becoming a requirement by more and more regulators and supply-chain partners as they strive to take advantage of the benefits of machine-readable coding of products and components. Increasingly, barcodes are being used in dispensing operations to reduce errors. Therefore, it is critical that the barcodes on packaging components are correct; any errors would normally lead to an immediate recall.

Like barcode creation tools, barcode verification software tools are available in a number of different formats to "read" a barcode contained in an artwork and report on its specification and information content. These tools can be used in the proofreading activity to verify that the barcodes on the artwork are correct.

There are also physical scanners that can be used to read barcodes and some operations use these as well as software components. These scanners come in two types. Those intended for supply-chain type scanning duties will simply indicate the information contained in the barcode, e.g. the product code. More sophisticated laboratory style scanners, or verifiers as they are often known, effectively emulate the software tools discussed and report not only on the information content of the barcodes, but also on the specification of the barcodes.

One further thing the laboratory style barcode verifier can do that the other tools cannot do is to verify the print quality of the barcode. There are internationally recognised standards that address different aspects of the print quality of barcodes and set standards for them: aspects such as contrast ratio, bar or spot size, linearity etc. Often, regulations defining the use of the barcodes will also define minimum quality standards for the printing of these barcodes and it is these verifiers that are required to check these quality standards are being met. Earlier, we touched briefly on the use of print press electronic inspection tools; it is possible for some of these tools to be equipped with barcode reading tools as well.

Braille verification tools

In the same way that barcodes need to be verified, so does Braille. Unlike barcodes, the verification of Braille on an electronic artwork is done in a very different way from the verification performed on a physical packaging component. The primary reason for this is the way in which the Braille is represented in the artwork compared to the physical component. In the electronic artwork, Braille is invariably represented by a matrix of coloured dots. On a printed packaging component, Braille is represented by embossed raised impressions in the substrate of the component.

As with barcodes, Braille verification software tools exist which "read" the Braille and report on its contents. These tools will normally require the Braille to be on a separate layer of the artwork drawing to allow them to distinguish the Braille from the often superimposed artwork graphics.

Tools for reading Braille on printed packaging components are very different in the way they work. The main issue is that there is no ink to read, only variations in the surface height of the packaging component substrate. For this reason, the tools often use distance measuring technology such as physical probes and/or laser distance measures to create a 3D map of the Braille indentations. Once the 3D map is created, the software then reads and reports the Braille contents in a similar way to the software tools. This clearly requires the physical sample of the artwork to be presented to the equipment in a controlled manner.

As with barcode verifiers, the physical Braille readers allow for the quality of the Braille embossing to be assessed. Most Braille requirements specify minimum quality standards for the Braille embossing and this can be verified by this technology.

Reporting

Before leaving the topic of proofreading tools we will say a little about the reports that can be produced by these tools, as they are important parts of the audit trail of artwork development.

Wherever possible, we would recommend recording the reports of the proofreading tools in the audit trail of the artwork creation. This provides an audit trail that the checks have been done, as well as providing potentially valuable information when investigating issues.

For both text and graphical comparison tools, wherever discrepancies are found a good tool should provide the operator with the opportunity to accept the discrepancies as acceptable and report them accordingly in the resulting proofreading report.

Integration

It is very easy to create gross errors in the process by selecting the wrong documents or by storing reports in the wrong places. Therefore, integration of proofreading tools with the document management system should be considered to help eliminate these risks.

Planning and workflow

As discussed earlier, the ability to direct the flow of tasks and route work from one operator to another can bring significant benefit to the overall artwork capability. Given this and the fact that there is much material available which describes the generic capabilities of these tools, we will restrict this discussion to those aspects specific to artwork management.

Before going any further, it is worth taking a few moments to define what we mean by workflow technology, as it is a term used to describe a number of different capabilities. For the purpose of this discussion we will take workflow to include:

- The ability to define a specific sequence of tasks for each artwork creation or change.
- The ability to define and control the relationships between tasks as one would in a project plan, e.g. finish-start relationships.
- The ability to define target dates for the completion of individual tasks.
- The ability to assign tasks to specific individuals and route work to them.
- The ability of users to see what tasks they have to do now and in future.
- The facility to report on the current status and past performance of activities.
- Provision of an audit trail for all activities.

Enhancements to the basic workflow capability that are particularly pertinent to workflow can include the following capabilities:

- The ability to create individual artwork change plans from template plans as many artwork plans tend to be very similar.
- The ability to make critical process control points mandatory when setting up an individual artwork change workflow, therefore using the software to mandate the key steps in the process.
- The ability to create suggested dates for activities based on standard lead times, allowing these to be manually adjusted as required.
- The ability to assign tasks to a work group and then, when appropriate, for the tasks to be assigned to, or picked up by specific users

within the work group. This aids in maximising the utilisation of key resource groups.

- The ability to manage task delegation and resource assignment switching. This allows issues such as absence to be managed effectively and also for workload balancing to be carried out more easily.
- The provision of a "My Task" web portal where users and managers can see a holistic picture of the activities and performance measures pertinent to them.
- The facility to alert users of tasks that are due for completion in the near future, thus helping to ensure that work is done on time.
- The facility to alert users and managers of tasks that are falling or have fallen behind schedule in real time. This allows issues to be dealt with in a timely manner.
- Integration of the alerting capabilities into other corporate communication tools such as email, SMS messaging and personal intranet pages. This ensures that users get the alerts in a manner that suits their way of working.
- The ability to group individual changes together to relate to parent change triggers or programmes of change activity. This aids significantly in searching and reporting.
- The ability to forecast workload on individuals and teams to allow active workload balancing. The facility to run scenarios can also be useful here.
- The ability to allow third parties directly into the workflow system. This eliminates the need to specially manage third party contributors to the overall process.
- The facility to segregate information visibility based on user group membership. This may be required for confidentiality reasons and

therefore, as an example, be a necessary pre-requisite of allowing third parties into the system.

- Audit trail and reporting of individual activity and performance.
- Links to other systems to avoid low-value-add operator tasks. For example, linking the completion of key milestones in the process to systems such as the corporate ERP system.

The validation approach to the workflow technology is always a topic of debate. On one side of the discussion is the approach that the purpose for the workflow system is to assist with the management of work execution and does not impact GxP information in and of itself. In this case, so long as it is ensured that these arguments are true in the design and use of the system, then it can be argued that the workflow system does not need to be validated. However, if at any point the system is used for GxP decision-making, information provision, or ensuring process adherence, then the system would probably need to be validated.

Forecasting and budgeting

In its most basic form, support for the forecasting and budgeting process can be provided by spreadsheets; indeed we have run one of the world's largest artwork management capability forecasting on spreadsheets.

As we pointed out in the supporting processes chapter on this topic, the forecast needs to be compiled using information that will indicate how many artwork changes there are likely to be in a given period. This means that sales volume forecasts are unlikely to be of any use, as in most business settings these give a poor indication of the artwork change volume. Therefore, linking to information sources in your sales and operations planning systems is unlikely to be of much benefit.

To aid the "planner with a spreadsheet" approach, one may consider setting up a simple portal to allow key people from across the organisation to enter forecast information. This can then be consolidated and challenged as part of the forecast building process.

Capacity and constraint modelling of artwork capabilities should be able to be handled by spreadsheet formulas for the most part. Depending upon the resource that is being assessed, you may find it useful to analyse the information at different levels of granularity, which in turn informs on the level of accuracy required in the forecasts provided. As discussed in the supporting processes chapter, KPIs should be used to derive simple relationships between forecast data, artwork changes, capacity and cost of the various elements of the artwork service.

What-if type scenario analysis can also normally be accommodated by simple spreadsheet analysis, or using the scenario tools available in more sophisticated spreadsheet applications.

The ability to cut the forecast data by product, geography, dose form, component type and other ways may prove to be useful when reviewing information with different stakeholder groups, or to analyse the key drivers for changes in the forecast.

Change control and authorisation

Most companies will have one or more change control systems to help ensure that change which could affect product quality and patient safety is managed effectively. Artwork creation and changes are clearly a type of change which impacts product quality and patient safety and therefore it would naturally follow that artwork changes are somehow incorporated into the quality change control system.

At one end of this spectrum, a company may have no appropriate vehicle to manage all artwork changes from a quality change control perspective and may therefore choose to provide the whole capability within the artwork process. In this case, capabilities such as the workflow tool may be used to ensure that key control gates are passed through before changes happen.

At the other end of the scale, where a comprehensive quality change control system exists, it may only be necessary to reference artwork changes and perhaps manage key control points in the quality change control system.

Regardless of how this is achieved, there are a number of matters that need to be considered in the design of the solution:

- How will the hierarchy of a change trigger be linked through country and product impacts all the way down to individual artwork component changes?

- How will shared components and/or packs be managed?

- How will the company report on such statutory reporting requirements such as the successful completion of critical safety changes?

- Should templates be used for the creation of artwork-related change control entries?

- Should there be automatic flagging of change impact to components in other systems?

- Should there be automatic links from the change control system to any artwork workflow systems?

Clearly, a quality change control management system should be validated.

Bill of materials, documents and pack catalogue

One area of information that is necessary for the successful execution of artwork processes is that of bill of materials (BOMs) and documents (BODs). This is particularly important when performing impact assessments on change triggers to ensure that all affected products, SKUs, components and their associated documentation are identified.

It is quite possible that your organisation already has these capabilities in place within their ERP or other corporate systems. If not, you may find that the appropriate data sets need to be created and managed as part of the artwork capability. This decision needs to be made carefully by an organisation, as it has many implications on the control of components beyond artwork.

Some of the fundamental requirements as they apply to the artwork process are listed below:

- Relationship database of components and SKUs to support where-used assessments.
- Single global numbering system (system generated) for SKU, component and file.
- Version number up-versioning built into system functionality.
- Search and filtering on data fields.
- Status of components visible.
- Defined and controlled data sources.
- Compliant SKU, component and file numbering systems managed at a local level by robust business rules.
- Single data sets across all company systems.
- Corporate BOM capability, including all third party BOMs.
- Ability to create relationships between components to support shared components.

Depending on the degree of sophistication of other tools in your artwork technology suite, some or all of the above capabilities may be able to be built into those tools. Alternatively, you may decide to provide a fully integrated technology suite in the form of tools such as product lifecycle management systems. At the other end of the scale, simple databases may suffice for less complex operations.

Because these systems are often used to identify the impact of a change trigger and provide other quality critical information, they are normally validated.

Performance management tools

We covered some of the performance management capabilities when we discussed planning and workflow tools, which are mainly focussed on real-time performance management. In this section we will focus on the performance management capabilities required at the department or function management level to manage longer term performance issues.

Fundamentally, performance management relies on performance measures, principally key performance indicators (KPIs), to make sense of the current and historical performance of the operation and make management decisions accordingly. Depending on the elements of technology that are put in place, the opportunity for the automated collection of this information varies greatly.

At its simplest level, KPI data gathering and reporting can be done on a spreadsheet or in a simple database driven by one or more individuals. In small operations this is a perfectly acceptable way of proceeding.

As the scale of operations increases, so does the value in having KPI data gathering automated. Clearly, if tools such as workflow management and electronic document management technology are put in place, then much of the data required for KPI reporting can be generated automatically.

In all cases we would recommend having the ability to generate reports which include performance against static targets, averages and other statistical measures and graphical trending. With this combination of capabilities, management will have an excellent tool-set to understand and drive performance improvement.

As systems get more sophisticated and integrated, the ability to "drill down" into the performance data in real time can also become very useful in pinpointing more specific causes driving performance trends.

For KPIs to be meaningful to a particular management team, it is preferable to be able to divide at least some of the KPI data up into subgroups which align with the particular management team who are looking at the data.

Finally, providing visibility of performance information to a broad audience is considered by many to help improve overall organisation performance. Many KPI reporting tools available today will create web-based performance dashboards which are accessible to a broad audience across the business.

To this latter point, you may well find that your organisation already has an established performance data gathering and reporting capability in place. In this case, it is a matter of interfacing the appropriate data to the system and configuring it accordingly.

When considering the validation requirements of performance management systems, many would argue that they are not directly associated with GxP decision-making or information creation and therefore do not need to be validated.

Translation management

In its simplest form, the management of translations can be achieved within a normal document management framework by creating individual documents for each required translation. This can be very effective where only a small number of translations need to be managed.

Once the volume of translations starts to increase, the use of more sophisticated translation management tools needs to be considered. Typically these tools manage translations by breaking documents down into individual phrases and managing translations for each phrase individually. Translated documents can then be built up by using already translated phrases. Each phrase translation is managed as a mini-document, going through the typical development cycle to ensure that only phrases which have been appropriately approved are used in final documents.

The benefits which accrue from the use of these types of translation tools include:

- Reduced translation errors.
- Reduced cost and lead time.
- Development of company and industry standard translations.
- Improved consistency of phraseology across products and markets.

There are opportunities for an organisation to select a translation agency that provides these tools as part of their service.

Technology downsides

Clearly, it is not all up-side, as anyone who has worked with technology knows. Technology comes at a cost, both initially and on an ongoing basis. It requires whole new skill sets to set up and manage effectively. If organisations are not careful, the benefits a technology provides can be quickly overtaken by these costs. This is particularly true in the validated world of a GxP artwork process. Some examples of these costs include:

- Initial costs to select solutions, configure, test and implement them.

- Ongoing maintenance costs of systems and their associated infrastructure.
- Replacement or upgrade costs as the technology becomes redundant.
- Education and training requirements to use the technology.
- Support capabilities to ensure users can continue to operate the systems.
- User access management capabilities.

Whilst technology provides the benefits to users, it also comes with what are often perceived as disadvantages. Tools are often complex or not ideal to use. Systems force people to fit their activities rigidly into the way the technology demands. There are fewer opportunities for individuals to optimise the way they work to suit themselves.

Far too often, technology is also seen as a solution for issues that are otherwise difficult to solve, such as organisation- and people-related issues. For example, be wary of technology proposals which are pitched as the solution to issues such as different functions not participating together effectively in the process. Technology can help in these situations but is rarely, if ever, the panacea people hope it will be.

Weighing the relative benefits of technology against full cost is also difficult. An organisation has many opportunities to introduce technology in this area to make improvements. Careful prioritisation needs to take place to ensure that the next steps you will take along the technology journey will bring the most net benefit, given your current situation.

Another aspect that must not be forgotten, particularly when introducing technology that manages the artwork process, is that it brings much more visibility to the organisation about what individuals are doing and how they are performing. With technology such as workflow tools, it may be possible for the first time for anyone to see what we are doing and how we are performing. In an ideal world, in a supporting culture, this will be seen in a positive light, providing opportunity for improvement. Unfortunately, in many organisations it can easily be seen by operators as a threat and can be used by management as a weapon. This can lead to a great deal of counterproductive behaviour and resistance.

9
Outsourcing

Outsourcing, as we will use the term in this book, refers to hiring an external organisation to provide all, or part of a business capability. You already very likely outsource part of the overall packaging component supply process by employing an external printing company to make and print the majority of your printed packaging components. This is outsourcing in the context that we mean it here.

For the purposes of this book, we are excluding the hiring of individuals to fulfil particular roles on a temporary or contract basis from the outsourcing discussion.

We will leave any discussions about hiring organisations to assist you in delivering any transformation of your artwork capabilities to the making it happen chapter.

Before we go any further, let us make it clear that it is highly unlikely, if not impossible, that any organisation will be able to outsource the whole of the artwork capability. As we have discussed several times, the cross-functional, global and cross-organisation nature of the process means that, at minimum, some aspects of the process are highly likely to remain in-house.

Considerations for outsourcing

So why would you outsource some or all of the artwork capability? We will look at the following drivers for outsourcing as a way of helping you answer this question:

- Does the activity bring competitive advantage?
- What is the risk that failure of the process presents?
- Scale of activities.
- What are the external opportunities?
- What internal capabilities are available?

The basic concept of competitive advantage, when considering outsourcing, is that if a capability is considered to give significant competitive advantage, then there is value in becoming excellent at it. Furthermore, there is value in not letting your competitors know exactly how you do all of it. Let us remind ourselves of some of the things that an artwork capability does and you can then consider if any of these give your business a competitive advantage.

Creating new artwork is synonymous with developing new products and your business may be based on launching a significant number of new products each year. For example, this may be as a result of a strong product development pipeline, the short lifecycle of your products, or perhaps it is due to an aggressive acquisition plan. In any of these situations, your ability to develop new packaging designs successfully, in line with external regulatory requirements, on time and to a high quality, is imperative to the success of your business strategy. We have worked with organisations which have made an art out of launching products within hours of regulatory approval and this requires certain aspects of an artwork capability to be "best in class".

If your business has a strategic imperative to expand products into new markets rapidly, then – in a similar way to launching new products – your business will not meet its strategic objectives unless it is capable of successfully delivering the associated packaging designs on time and to the right quality. You will remember from the first chapter of this book that this is a position that many businesses find themselves in today. Again, this requires certain aspects of your artwork capability to be "best in class".

Perhaps you operate in a business where your products change or evolve frequently. In this situation, you will frequently find that the artwork changes are on the critical path of product delivery.

Increasingly, governments and other large purchasing organisations are doing business through tenders. Often tenders have specific packaging requirements which must be met and therefore new or revised packaging is required for each new tender. Often, as part of the tender submission, a company is required to submit mock-up artwork designs. Furthermore, the nature of this style of business often means that the time between being awarded a tender contract and the product delivery date is very short. This all puts a high quality, fast turn-around artwork capability at the forefront of delivering this business.

If, after considering the above discussion on competitive advantage, you may conclude that there are particular areas where an artwork capability will give your organisation significant value, then this should influence your thinking when considering outsourcing. That is not to say that it should preclude outsourcing options; indeed it may drive you to consider outsourcing certain activities to best in class suppliers. However, it should also ensure you consider the competitive implications of outsourcing decisions.

Another factor to take into consideration when considering outsourcing is the consequences of failure of the process. Clearly, where artwork is concerned, the consequences of a single failure to the pharma company leading to a recall can be significant: loss of sales; direct costs; management distraction; brand reputation damage; increasing external regulator scrutiny, to name the main implications. The consequences of failure of the process to those performing outsourcing services are likely to be much less severe, unless some form of punitive damages clause is included in the particular outsourcing contract. For this reason, a pharmaceutical company should carefully consider the overall design of the artwork capability, any outsourcing partner's role and where the key quality checks and decisions are being made.

The scale of your operation also has an impact on outsourcing decisions. For small organisations in particular, there is an interesting dilemma that needs to be considered carefully. On the one hand, as a small organisation you are unlikely to have the critical mass to enable you to develop an adequate internal capability for certain aspects of the overall artwork capability. As with all business activities, it should never be underestimated just how difficult it is to do any particular activity well. On the other hand, as a small purchaser you are unlikely to have a great deal of leverage with potential suppliers.

Larger organisations have the luxury of scale to allow them to consider keeping more of the required capabilities in-house.

Figure 9.1 looks at the different aspects of an artwork capability and examines some of the outsourcing opportunities that were available at the time of writing this book.

A final thought about considering outsourcing opportunities is to review the capabilities you may already have within your organisation. It may well be the case, particularly in larger organisations, that different parts of your organisation already have all or part of the capability you require. Either they will do the activity internally, or they may have already sorted out external service providers. We would offer a word of caution here when considering using internal

Capability Area	Opportunity	Comments
Approved text creation and negotiation with external regulators	Limited.	Needs to be managed directly by the pharmaceutical company, its joint venture partner or the local marketing authorisation holder.
Translation of regional text	Translation houses can provide this service.	Needs to have an overall quality approval process associated with the service managed by the pharmaceutical company.
Creation of artwork	Specialist artwork providers can provide this capability. 3rd party packing suppliers will sometimes provide this service.	
Proof reading	Specialist artwork providers can proof read their work.	The acceptability of this as the only proof read in the process needs to be carefully considered.
Packaging technology expertise input	Specialist service providers and/or packaging component suppliers can provide this capability.	
Pre-press and printer proof creation	Three principal options: – Eliminate by using print- ready artwork process – Packaging component suppliers – Artwork providers	The industry is beginning to move to eliminating this step
Printing components	Norm is to get 3rd parties to provide printed packaging components.	
Packaging component quality checks	Can move to qualified packaging component suppliers to eliminate many aspects of this. 3rd party packaging companies will provide this service as part of their overall packing service.	Normal to have a final release of product which involves quality checking by a qualified person acting on behalf of the pharmaceutical company as a final check.
Managing artwork changes across the pharmaceutical company's organisation and supplier base	Can be outsourced to third parties so long as they are empowered to carry out this activity by the impacted stakeholders.	Often managed by internal people.

Figure 9.1 Artwork capability outsourcing opportunities

services. Firstly, you need to refer to our discussion earlier in the book around the topic of organisational boundaries and governance to ensure that sharing of an internal service would be sensible. Secondly, just because another part of your organisation already has an external service provider in place, or provides a service themselves, does not mean that it will be acceptable for your needs. We have experienced several occasions, both ourselves and with our clients, when services provided by, or arranged by, other parts of the organisation have caused many issues because they were not what was required. However difficult it may be politically or otherwise, we suggest that you use full due diligence in deciding if an internal capability or internally managed service is acceptable.

The final point we would make about considering external service providers as a way to provide parts of the artwork capability is that outsourcing will not magically fix an otherwise broken business process. If anything, it will make the situation considerably worse, make issues very visible and give you a very difficult interface between your organisation and the external service provider to manage on a day-by-day basis. If your current process is broken you really only have two choices: fix it first and then outsource the appropriate parts, or redesign it and implement the new process with the relevant new external service partners.

Assessing external service providers

There has been much written on the subject of service provider selection and we do not aim to repeat a lot of that knowledge base here. What we will cover here are the aspects of service provider selection that have proven to us to be important when it comes to artwork capability outsourcing.

The basic areas that need to be looked at when considering any outsource partner, and artwork capabilities are no exception, are:

- Ability to deliver the required service in the short term.
- Quality compliance.
- Service level.
- Quality management system.
- Cost and stability.
- Strategic fit.

The starting point for any external service provision is to define exactly what you want the external service provider to do for you. This definition is normally based on a detailed business process design and a consideration of the points at which the service provider will interface with your process. Having defined what you want the service provider to do, you need to document it, explain it to the potential service providers and check for understanding. Our recommendation would be to visit the potential service providers and talk through their understanding of the service provision with them, seeing where the service will be carried out, how they intend to do it and discussing it with operating staff who will be doing it. This will do a number of things, all essential to a quality decision:

- Allow you to assess their capabilities thoroughly.
- Allow both parties to clarify their understanding of the requirements.
- Allow you to assess the way they work and their culture.
- Allow you to start building the relationships you will need should they be successful.
- Allow the service provider to adjust their proposal if any significant change is required.

In parallel with the above, you need to consider what quality control activities they need to have in place to ensure the quality of what they deliver. Depending on what they receive to start their process, it is also worth considering what quality control activities they might need to undertake to check any inputs they receive from you as well.

Along with the quality control activities, there needs to be an assessment of the potential service provider's overall quality management system. It is almost certainly the case that you will want the service provider to demonstrate a quality management system that meets pharmaceutical quality standards.

The cost of the service is always going to be important, no matter how large or small your organisation. When comparing costs of potential providers it is important to be able to compare like with like. Fortunately, when it comes to artwork-related services, much can be defined and contracted for on a rate card basis. For those of you not familiar with this term, we mean that the service can be defined as a number of deliverables and that a price can be defined for each deliverable. For example, an artwork provider can quote for the price to deliver a simple single language leaflet. If you define the structure of your rate card before asking for prices, then comparing the responses and modelling the full comparative costs will be much easier.

The long-term stability and viability of the service provider should also be of keen interest to a customer if a long-term relationship is intended, not least because the cost of switching suppliers is always very high. Your procurement team should be able to assist in the assessment of this aspect of the selection.

When we talk about strategic fit, we are really talking about three things. Firstly, if you perform the service provider selection well, you are going to have a relationship with the service provider for a number of years. Therefore, it is important to select an organisation with whom you are confident that you and your people are able to work. This can be quite an intangible decision criterion, only some of which can be understood through the response to questions that a potential provider may supply. As mentioned above, we would recommend using the selection process to get to know key people in the potential service provider, visit them in their own operation and make an overall judgement on this topic from all of those inputs.

Secondly, you need to understand if the strategic direction of their organisation fits with your strategic direction for the service. It is no good getting into a relationship with a service provider only to discover that they are planning to exit your segment of the business. Likewise, if you value a small niche service provider, it is no good entering into a relationship with an organisation that has just taken on venture capital to grow aggressively.

Supplier selection approach

An approach to the assessment of artwork service providers that has proved successful involves the following steps:

- Define and agree selection method with steering team.

- Create a shortlist of potential service providers. A request for information (RFI) may be necessary here if you do not have access to knowledge about potential suppliers.

- Prepare a request for proposal (RFP) and quotation (RFQ) document and send to shortlisted providers and create the supplier selection criteria and weighting.

- Receive and analyse RFP/Qs.

- Visit service providers with a small team of your people.

- Hold a selection meeting with the team that performed the visits and make a supplier recommendation.

- Present findings and recommendation to steering team for final decision.

The ultimate decision about which service provider(s) to select would normally be made by the steering team in many organisations. In order to ensure their buy-in to any recommendation, an important first step is to agree the approach to the supplier selection with them. In this way, when you present the recommendation at the end of the process, you will only be having a debate about which supplier(s) to select and not how the decision is being made.

Creating a shortlist of potential suppliers can be difficult for some areas of artwork capability supply and you may need to seek external help in doing this.

This selection process is normally led by the procurement department, but they will need support from the business to ensure comprehensive requirements are defined and assessed.

We have often witnessed supplier selection processes that involve what seem like endless rounds of written communication to clarify questions and points in the RFP responses between the customer and the potential suppliers. This has always struck us as a very inefficient way of doing things, often resulting in a selection based on the service provider who could stay the course, rather than necessarily provide the best service. This is why we recommend shortcutting this process and going straight from a read through of the RFPs to a site visit.

Ideally the visit should involve a small team of people from your organisation who are capable of covering off all the bases in one hit: assessment of the service provision and quality control; assessment of the quality management system; assessment of the company and their people; assessment of their IT capabilities (if appropriate). This will probably mean that you need an artwork specialist person, a quality audit person, possibly an IT specialist and a procurement specialist. The visit can be based on the structure of the RFP and should aim to ensure a complete understanding is sought for all RFP questions. The visit should also include a tour of the service provider's facility, preferably following a typical job, and include conversations with the individuals who carry out the service tasks along the way. This will ensure your team gets a thorough understanding of their operation, service offering and culture. It also provides an excellent way to cut through the sales pitch that might otherwise be the only thing you see.

To ensure consistency during the assessment and selection of suppliers, we would always recommend that the same team visit all the suppliers. Ideally, the team should also be involved in the assessment of the RFP/Qs prior to the visits. If this is not possible, then the team need to be thoroughly briefed on the RFP/Q analysis prior to each visit.

Establishing supplier contracts

Once selected, the contractual agreement about what a service provider is going to do for you will normally be captured in three key document elements. They may all be part of the same document or may be different documents; this often largely depends on current practice and authorship practices in an organisation:

- Contract.
- Technical and quality agreement (TQA).
- Service level agreement (SLA).

The contract is normally the parent document, covering all the legalese associated with the service. It is often highly standard in nature and references the other documents which describe what the service provider is going to do. The cost of the service is normally included in the main contract. This document is normally owned by the procurement and/or legal function.

The technical and quality agreement would normally describe in detail the service activities that the service provider will carry out and the interface of that service with your and any other relevant organisations. If appropriate, it should also reference any special tools or IT systems they will use. Finally, it should spell out the quality standards and quality management system that the supplier will use in delivering the service. This document is normally jointly owned by the artwork function and the quality function in the customer's organisation.

From an artwork-specific perspective, it is worth remembering that many of the risks are in the detail. Therefore, the exact way that the service provider performs their tasks and the tools they use are key to a quality output. Therefore, the TQA should spell out all the elements which are essential to the quality result of the artwork. It should also define the circumstances under which any change or variation to these agreements are to be allowed and how these deviations will be approved. This detailed definition of how the service provider will carry out their tasks must be balanced with ensuring that the service provider has the ability to offer the most appropriate service to the customer. We suggest this is best achieved by ensuring that the service provider leads the recommendation of how to provide the service and that changes are only made with both parties' consent when there is real value in doing so.

Finally, the service level agreement should primarily deal with agreements associated with the response times that the service provider will meet and the capacity they will ensure is available to the customer. This document is normally jointly owned by the procurement function and the artwork function in a customer organisation.

Let us take a more detailed look at an example to illustrate some of the issues pertinent to artwork service provision. We will use an artwork creation studio SLA as an example. This service provider will take an approved brief and create the artwork in accordance with the brief and hand it back to the customer. Firstly, it is normal to break down the service deliverables into some sensible sub-groups, normally following the effort and time involved in creating them. In this case, the SLA may define service performance for labels, carton artworks and leaflet artworks separately. Furthermore, it may differentiate complex artworks from simple artworks; for example, single language leaflets from multi-language leaflets. This breakdown would also normally be mirrored in a schedule of rates in the contract. Having identified a breakdown of the service deliverables, the SLA should define the response times for each deliverable. Here we should recognise that there is sometimes a need, and a resulting cost, for rush work to be carried out. It is therefore normal to define a normal turnaround time for the different deliverables and a rush turnaround time. The agreement about the split of normal to rush work can be dealt with in a number of ways. An agreement may be reached to define a maximum proportion of the work in a period which is rush type. Alternatively, an increased price may be agreed for rush work with no cap on the proportion. The important thing here is that both parties agree on the model up front.

Managing service providers

In managing artwork service providers, it must be recognised that you are managing a long-term professional relationship. Lasting relationships typically require a number of things to survive:

- Mutual benefit from the relationship.

- Ongoing commitment and effort to make the relationship work from both sides.

- Mutual respect.

The initial agreement needs to be set up in a way that ensures both parties feel they are getting some significant benefit out of the relationship. We would suggest that a relationship based on the customer getting constant wins at the expense of the supplier always feeling like they lose may feel good for the customer for a while, but the relationship is likely to be very difficult to manage and will end at the first opportunity that the supplier gets.

Like all relationships, it takes effort and commitment to make it work. Therefore, the management of the relationship should create the opportunities to put in this effort on both sides. Some techniques that can be used to make this happen include:

- Agreeing the day-to-day communication lines for normal work and issue escalation.
- Establishing regular service review meetings.
- Sharing of objectives from each party for the upcoming period.
- Sharing of workload forecasts from the customer.
- Sharing of capacity issues from the service provider.
- Keeping track of performance of both parties through the use of KPIs.
- Actively identifying and managing risks and issues in a way that tries to accommodate the needs of both parties.
- Mutually agreeing improvement activities and priorities for both parties.
- Defining mutually acceptable behaviour of staff on both sides of the relationship and actively managing unacceptable behaviour when it occurs.

- Celebrating success, both at the organisation level and the individual level.

- Personal visits by the customer to the service provider, including discussions with the service provider team, management and operations staff.

- Performing periodic quality audits of the supplier and incorporating any corrective action into the overall improvement plan.

Outsourcing doesn't come for free

If you were under any illusion before reading this chapter that outsourcing was an easy option, we hope that you now see that it is not. That is not to say in any way that it is not a very worthwhile approach; it is just important to recognise that it needs different effort and skills to achieve successful results. You should also recognise that whilst you can outsource activities, you cannot outsource accountability.

As a customer, you need to recognise that successfully establishing and managing these relationships requires you to invest in a different set of skills and resources that requires constant and ongoing effort to make it work successfully.

As a last thought on the topic, we would re-emphasise a basic mistake that is repeated time and time again: don't outsource a process that does not work, it will not solve the problem and will likely make things a lot worse.

10
Future developments

There is one thing for certain in discussing potential future developments: any predictions that we make are bound to be wrong to a greater or lesser extent. Also, no sooner than we put pen to paper, the information we provide starts to become out-of-date. Nevertheless, in this chapter we discuss some of the potential developments that may impact artwork capabilities in future and we hope that these thoughts help you navigate your way through the changes to come.

We will cover the following potential developments in this chapter:

- Legislation development.
- Further enhancements of today's tools.
- XML and standardised regulatory submission management.
- On-line printing.
- Web and off-pack product and patient information.
- Content management systems.
- Automated artwork creation.
- The Cloud.
- Seamless virtual artwork capabilities.

- Supply base development.

Legislation developments

As we have already discussed, the legislative and enforcement environment surrounding artwork capabilities is likely to become more stringent over time. The consequence of this is that the capabilities which were considered acceptable yesterday will no longer be appropriate in future.

At the same time, new legislation regarding product packaging and labelling is evolving all the time as regulators around the world strive to improve product and patient safety. Examples of such legislation include:

- The requirement for use-testing of product designs as part of the product approval process.
- Increased child-resistance requirements.
- Measures aimed at reducing counterfeiting.
- Measures aimed at reducing fraud.
- Measures aimed at improving compliance and persistence.
- More inclusive packaging design for visually and physically impaired users.

In recent years, internet-based pharmacies have come into existence. An immediate impact of the rapid expansion of this distribution channel has been the move by some pharmaceutical companies to introduce packaging specifically optimised for postal delivery. At the time of writing this book, little effective legislation seems to exist to manage these internet operations and there is much press reporting of product and patient safety related issues arising in this model, not least from the use of the channel to introduce counterfeit drugs. Undoubtedly, as effective legislation for the management of this channel is introduced, it will have an impact on packaging design requirements.

Organisations need to put in place mechanisms to monitor the legislative environment relating to packaging and labelling capabilities and ensure that they take timely action to meet the emerging legislation. One aspect of this strategy should include the potential to get involved in the development of emerging legislation to help ensure that it is actionable and effective.

Further enhancements of today's tools

Technology is an ever-evolving area and this will undoubtedly lead to the further development and level of sophistication of the tools that are available to artwork capabilities today.

It is without question that applications such as proofreading tools will continue to develop over time to eliminate their current limitations and provide new and more useful features.

A trend that we see all the time with technology is the integration of specialist capabilities into more general tools. This trend is likely to mean that today's specialist barcode-generation tools become standard features of tomorrow's artwork creation suites.

Another trend that we observe is a focus on better integration of tools. Today, if you want a fully integrated suite of tools to support the artwork capability, significant investment is the only option. Either you opt to spend resources and money interfacing a number of separate tools to achieve the result, or you invest in an already integrated suite based on technology such as product lifecycle management suites. The move to web-based services and what has recently been termed the Cloud (discussed further below) will increasingly mean that it is easier to group together different tools into virtual single applications at significantly lower cost than is possible today.

XML and standardised regulatory submission management

Much has been talked about in recent years about the use of XML technology to help structure regulatory submission documentation including artwork source text documents. The primary drivers here appear to be improving the quality of regulatory approvals and making the process faster and more efficient for all concerned.

The underlying principle of these moves is to break key regulatory text into its constituent elements and re-use those elements consistently wherever they are needed, in whatever language they are needed. In this way, the elements can be managed once and not, as is currently the case, in every place that they are used. This is a practice that has been present in other areas for some considerable time; as an example, the construction of material safety data sheets follows these principles.

Such developments exchange one set of challenges for another. In this case, the goal is to be able to improve effectiveness by managing the building blocks of text once and this is very appealing. However, in achieving this, a number of other new aspects of regulatory text management will need to be explicitly standardised and managed. For example, standards need to be agreed on the structure and content rules for each element of text. This is a significant task in its own right. Furthermore, whereas today most regulatory text is created and managed with standard computer desktop tools such as Microsoft Word, in an XML world complex and dedicated applications will need to be used. This all adds a cost and management overhead to moving to this new technology.

XML technology has become synonymous with these developments simply because it provides a relatively new and potentially effective framework for realising these aims. It is not the only technology that could achieve the end result, but it looks like a strong candidate today.

A number of years ago, the European Medicines Agency launched an initiative to develop XML-based technology to help manage the electronic submission and approval of products registered under the European central procedure regulations. Recently, the project was put on hold and it is unclear at this moment how this project will progress in future.

A number of other governments have shown public interest in developing this approach and it will be interesting to see how this promising area develops over the next decade.

We suggest that one topic that the pharmaceutical companies should address is that of standardisation across government legislation. As we have seen in areas such as product serialisation, the evolution of different requirements from different regulators can drive significant cost and complexity for pharmaceutical companies without adding significant benefit. As it is highly likely that pharmaceutical companies will pick up the majority of the costs of these technologies, it is in their interest to drive for optimal levels of standardisation in this area.

On-line printing

There are many ongoing developments in the area of printing packaging artwork on, or near to, the packaging production line, all of which have potential impact on the way in which packaging artwork is managed. Technologies such as on-line blister foil printing and on-line label and leaflet printing have the potential to move the end point of the artwork management process into packaging machinery and inspection systems on-line.

There are many business drivers which make this technology increasingly attractive, some of which include:

- Meeting legislation requiring batch- pack- or dose-specific marking.
- Reducing packaging lead times.
- Improving customer service for low demand through postponement and late customisation.
- Providing patient-specific packs.
- Reducing packaging component waste.

Traditionally, the only information to be printed on-line was batch-variable data such as batch/lot and expiry date. The batch-variable information is a necessary part of the batch production instruction and therefore presented no additional management issue. The specification of the actual printing, such as font, font size and location were often managed and restricted through the selection of printing equipment such as hot-foil coders. Furthermore, due to the nature of the way in which these simple technologies work, once set up for a batch, if there is any marking achieved on the pack there will be a very high degree of confidence that the right information has been printed. This in turn makes the verification of the printing a relatively simple and often manual activity.

Over recent years there has been a move to more sophisticated on-line printing technology. Regulations which require the printing of machine-readable batch-variable information and serialisation on packs, using items such as 2D data matrix codes, require much more sophisticated electronic printing technologies to be used at or near the packaging line. Several issues emerge from this with respect to artwork management:

- How is the information content to be included in this pack marking to be managed?
- How are the detailed specifications which define the codes and print location(s) to be managed?
- What print verification methods need to be used, given the significantly increased number of failure modes that this technology presents.

As we move to even more sophisticated technology which allows the printing of complete foils, cartons, labels and leaflets on-line, the problem only gets more complex, often involving the merging of content with templates in the printing machines.

The artwork capabilities that we have discussed in this book aim to ensure that all text and graphics that appear on finished packaging components is correct. However this information gets on to the pack, the overall quality requirement remains unchanged. Therefore, organisations need to consider how and where to manage this information as they move forward.

Web and off-pack patient information

A primary concern of regulatory agencies and pharmaceutical companies is to ensure that prescribers, administrators and patients have access to accurate and current information when they need it. The traditional method of achieving this is to rely on the information provided with the drug's packaging in the form of the various types of artwork. Whilst very successful, this method has a significant downside: the information was only current at the time the product was packaged. If new information is required to be shared, or errors are discovered in the information provided, the costly and, many would argue, not very effective product recall route needs to be taken.

By moving information off-pack, opportunities also arise to provide information in different formats. Non-profit organisations such as X-PIL are already providing patient information leaflet information in large text, audio and Braille formats for the visually impaired.

The internet provides the opportunity to make such information available to all stakeholders almost instantly. Changing this web-based information does not necessarily require the recall of physical product already in the supply-chain. There are also potential benefits in the reduced overall costs of managing electronic information when compared to today's paper-based information.

Furthermore, with the drive to improve compliance and persistence, more use is being made of a combination of on- and off-pack information and tools. Links to electronic information and tools from the pack becomes increasingly important. For example, the rise in popularity of the smart phone provides an opportunity to provide applications which can be triggered or fed with information from the packaging. Today this is achieved through the use of quick response (QR) matrix barcodes.

As with our discussion on XML content management, such a move to web-based information provision comes with a host of new risks and issues which need to be managed, all of which come with their own cost implications. A simple issue such as ensuring that the links contained in QR codes printed on packaging are always available for the life of the product is an example of the type of thing which needs to be carefully managed.

With these potential benefits in mind, the FDA launched an initiative to put information leaflets on the internet. At the time of writing this book the situation appears to be:

- Most texts are already on the internet.
- A number of issues need to be resolved, including:

- When is the tipping point that will enable leaflets to be removed from product packaging?

- How to manage ownership and liability issues as product and patient safety information is provided on new web-based systems.

- How to cater for visually impaired patients and other groups with specific needs.

It will be interesting to see how this initiative progresses over the coming years.

Content management

Content management systems (CMS) have emerged to help organisations manage the creation and use of information content across a complex array of electronic and physical media. Today, a number of translation agencies use CMS technology to manage the translations which end up on artwork. As patient and product safety critical information is provided on the web, there will be an opportunity to harmonise the technology which manages this information content.

Initially, patient and product safety critical information is likely to be dealt with in CMS as discrete "locked" documents. Even when dealt with in this limited way, all the issues we discussed in the technology chapter need to be addressed with respect to ensuring the user is presented with an accurate rendition of the information, be it on screen or from a printer.

As technologies evolve, we may well see the merger of many of the technology elements required by artwork capabilities with the functionality provided by CMS.

Automated artwork creation

The concept of automated artwork creation is straightforward: develop a standard template for an artwork; define the content of the artwork in a standard way; automatically populate the content into the artwork and adjust the format of the artwork and content as required, without the need for manual artwork operators. However, as with practically all automation endeavours, there are many detailed challenges to overcome to achieve a robust solution.

The potential direct benefits of this approach are the elimination of operator errors in the creation of the artwork, the reduced time taken to create the artwork and the reduction in resource costs in the artwork creation and verification process. The indirect benefits include things like improved brand consistency. Indeed, if one looks at where this technology is being initially adopted, it is in the world of fast-moving consumer goods (FMCG), where short product lifecycles and significant reliance on costly brand marketing drive a need for rapid packaging design development with rigorous adherence to brand image.

It will be clear by now, having discussed XML content management previously, that this move to automated artwork creation brings with it a different set of risks and issues that need to be managed, some of which include:

- The need to drive standardisation in artwork layout through the use of templates.

- The need to develop standards for the breakdown of text and graphical content to suit the artwork template format and to drive this into the functions that develop the original content.

- The need to develop and manage the IT technology to perform the automatic content creation in a manner that will eliminate the possibility of errors.

As with the move to XML regulatory submission management, the change management challenge of moving to this model will be significant. To be successful, the artwork content structure requirements that are necessary to enable the automated artwork build tools to work need to be driven right back into the structure of the artwork brief definition, and potentially even further back into the text creation processes. Clearly, if an organisation is already in a position of having a highly structured and standardised methodology for defining and building artwork, then the change journey is already partly complete.

At the time of writing, two forms of automated artwork creation are available in the marketplace. Firstly, there is the "full" model we have just described. The second is a much scaled-down version which simply automates the "flowing" of individual blocks of text into pre-defined areas of an artwork template. This latter option is considerably easier to achieve technically, but does not bring many of the potential benefits of the "full" model.

The Cloud

One broad definition of the Cloud is the delivery of hosted services over the internet. Cloud-based technologies provide the opportunity for service providers to provide a combination of web-based applications seamlessly and host them at low cost to the user. Often the pricing models associated with these service offerings are on a per use, or per user per month basis and the prices can be surprisingly low if economies of scale can be achieved.

By way of an example, there has been a move recently to the provision of basic office technology tools such as email, information repositories and corporate communication and collaboration tools to the Cloud. Companies such as Microsoft have developed service provider networks that provide companies of all sizes with access to all of these latest technologies at a fraction of the cost of providing and managing these capabilities for themselves. Furthermore, these services become largely variable business costs and free up precious capital to be invested in driving the business forward.

There is no reason why the Cloud model should not be applied to the tools which support artwork capabilities. This is particularly true now that technology providers are addressing the clunky user-experience issues related to traditional web-based applications. Whilst the nature of artwork activity means that the economies of scale are always going to be considerably smaller than, say, the provision of Cloud email service, economies of scale are nonetheless evident.

When considering Cloud technology solutions, all the regular IT outsourcing service considerations need to be taken into account including things such as validation requirements and data security.

This will be an interesting space to observe in the coming years and may provide the opportunity for companies to have access to much more sophisticated artwork technology tools than is currently economic for them to consider.

Seamless virtual artwork capabilities

As software capabilities evolve with technology advances such as the Cloud, the real opportunity for cost-effective, virtual artwork capabilities emerges. In this future world, the boundaries between organisations will become increasingly seamless and organisations will be able to construct virtual artwork capabilities by plugging and unplugging sub-capabilities with ease.

This journey has already started with the increasing use of tools which allow the seamless sharing of information across organisational boundaries and the direct involvement of third parties in business process work management technologies.

Currently, such models tend to be bespoke, being defined by the larger pharmaceutical companies or the more dominant suppliers. For seamless artwork capabilities to flourish, it will be necessary for the industry to develop standards for business processes, the exchange of information and the security measures that need to accompany these activities.

Supply base development

Finally, as suppliers take advantage of some of the developments discussed above, they are bound to expand and evolve their offerings, which in turn will present customers with different opportunities.

Currently, we would note two particular trends in the artwork capability supply base:

- The move for artwork creation studios to provide an increasing scope of services.
- The emergence of independent internet-based artwork management applications.

An organisation should put in place mechanisms to stay in touch with such developments in the supply base, allowing them actively to decide when such developments are suitable for adoption into their artwork capability model.

11
Making it happen

At this stage of the book you should have a good understanding of the elements of an artwork capability, but you are probably asking yourself a fundamental question: how do we go from where we are today to where we need to be? In this chapter we will discuss the improvement change journey, breaking the discussion down into four parts:

- Deciding what needs to be done.
- Designing the change activity.
- Making it happen.
- Who can help make it happen.

Ensuring sustainability of the result is dealt with in various ways in other chapters of the book – for example, thorough documentation of the process and supporting process to drive continuous improvement.

Deciding what needs to be done

Making improvement or change happen to any capability is analogous in concept to taking a journey. Before deciding how to go about the journey you need to know or understand two fundamental things: where you are now and where you want to go. In change management speak you may have heard these two points referred to as the as-is and the to-be or future state.

We would recommend a holistic approach to this activity, the result of which should be a clear understanding of all the areas of the artwork capability that need to be improved and in what priority. Taking a holistic approach may make the first step more difficult and time-consuming, but it will pay dividends in the medium term. All too often, we see a piecemeal approach to improvement activity definition which results in resources being dedicated to areas which do not deliver the best return for the organisation. Perhaps worse than that, the piecemeal improvement activities discredit the artwork organisation for not addressing what the stakeholders perceive as more important and/or pressing issues and could make performance worse in the short term through disjointed change activity.

Defining the steps

Having decided where you are and where you want to go, you need to decide how quickly you want to get there. This is partly driven by the need for getting to the destination and partly dictated by the transport options available to you. You may have a pressing need to resolve a particular issue that has arisen. Alternatively, you may have some longer-term goals in mind.

Conceptually, it is sensible to think of the change journey as a number of smaller trips, the sum of which results in the overall journey. Each trip will have different destinations, different imperatives and different options for getting there. The trick is to break down the overall change activity into sensible pieces and focus on delivering each piece in the appropriate timescale.

For most organisations, there are many improvements that could be made to their artwork capability. Furthermore, everyone sooner or later has to face the question: how far is far enough? Or to put the question another way: how good is good enough? Each element of a potential to-be will have different levels of performance and efficiency to which it can be delivered. Let's consider these as a sliding scale of potential solution options. At one side of the scale are solutions which meet basic compliance and control requirements; perhaps a paper-based solution is good enough to reach this level. At the other end of the scale are solutions which deliver best in class, or state-of-the-art solutions. In the latter case, the solution to the same to-be requirement may be highly integrated process and information technology solutions.

The good news is that, for most organisations, achieving best in class for all areas of an artwork capability is not necessary to meet your business's strategic objectives. Indeed, for many organisations, it is not necessary to achieve best in class in any of the artwork capabilities. This may be particularly true in the early stages of developing your artwork capability. As in many things, it is important to learn to walk before you can run. The possible exception to this analogy is that you may choose to get someone else to do the running for you, at least in part, which gives you the opportunity to transform elements of your performance very quickly. We talk more about this in the chapter on outsourcing.

Given that there are a number of end points to choose from for any given to-be capability, it is necessary to do some work to decide what level of sophistication is required in the to-be solution to meet the needs of the business. In doing this exercise, you would not be the first company to realise that it is probably sensible to take each capability through manageable stages of evolution. We have had some success in getting organisations to think of this evolution in terms of three levels:

Level 1 – Achieving compliance and control.

Level 2 – Becoming efficient and effective.

Level 3 – Becoming best in class.

You will no doubt find that there may be some areas of artwork capability that provide core strategic value to your business. In this case, you may decide that achieving level 2 or even level 3 quickly is very valuable to you and worth investing a great deal of resources to achieve. For other areas, it may be sufficient to achieve level 2 over a longer period of time. For what should be obvious reasons, we would always argue that it is necessary to achieve level 1 competence in all areas of the operation that have impact on patient and product safety.

If you have a specific issue at present, it is worth doing some root cause analysis to determine what the main reasons for failure are in the current situation. We would advise caution here. It is all too easy to embark on what quickly becomes a huge root cause data analysis exercise which results in little more than a room full of data, no better understanding of what the issues really are and an exhausted and fed-up team. We suggest a balance of data-driven analysis coupled with pragmatic judgement of experienced people.

Having understood the principal root causes of the current poor performance, you should be able to match these to solutions in your to-be. This then helps you group and prioritise a part of your improvement journey.

This approach can be repeated for each significant element of business performance improvement that is desired until you have a complete plan of when the different elements of your to-be need to be delivered. Clearly, this exercise is somewhat iterative as many of the elements of the to-be capability cannot be delivered in isolation.

Designing the change activity

By this stage in our hypothetical journey you will have defined a series of milestones in time, and for each milestone there will be a definition of what the desired to-be needs to look like. Now comes the task of designing the change activity, or projects to achieve this.

We are not about to launch into a full description of how to design projects as this could fill a book or more in its own right. We will, however, touch on some of the key aspects to consider when designing artwork projects.

The first thing to consider is your change management approach to any particular project. By change management, we mean the way in which you will interact with the various stakeholder groups to ensure a successful and sustainable project outcome. Artwork development is a highly people-driven process and people are therefore key to the success of any artwork change activity.

Contrast this for a moment with a production line operator's world. He or she operates the production line day in, day out, for almost the entire time he or she is at work. Furthermore, the operator goes to a purpose-built environment to carry out these tasks – the production line. In the world of the production operator, it is clear that they must turn up to the production line and execute the process more or less exactly as required. If they do not, there will be immediate and obvious consequences: the line will stop, the quality control inspection station will fail products, production output will drop and the list could go on. An intelligent operator doesn't need to be told that following the process at the right time is important; it is obvious. Furthermore, the management of the production process are all close by and can come and see what is happening on the production line any time they need to. The production management are also likely to be fairly empowered to make decisions about changes to the way in which tasks on the production line are carried out, so long as this does not adversely affect the product.

In the world of people executing an artwork process, things look very different. More often than not, they execute their part of the process in isolation. As discussed earlier in this book, the work comes to them at their desk and it is often only a small part of their daily job. Indeed, for some people located in small commercial sales operations, they may only execute the process once or twice a year. Furthermore, they are unlikely to have ever met many of the "team" who are responsible for making a particular artwork change happen. What is more, many members of the "team" will change each time they work on a different product or pack. If the person in question does not know exactly how to do their job and they do the best they can, potentially adverse results are highly unlikely to be immediately obvious.

Now consider the management of the artwork process. Because the artwork process is a multi-functional and often multi-company process, no single management team can have responsibility for either the full process or all the people executing it. Furthermore, there is the issue that the management team of the process do not have a production line to go and look at. It is very difficult to observe directly what is happening in the artwork process; the nearest thing the management team may have is the process KPIs and incident reports. All of this makes management of the ongoing process and any changes to it particularly challenging.

It is for the reasons above that the design of an artwork project needs to be done in a way that involves all impacted stakeholders. This will help to ensure that the people impacted by the change feel involved in the development of the new capability and understand the decisions that were taken during the design. It will also help to ensure their buy-in to it. Furthermore, to stand any chance of the process being sustainable, support processes also need to be put in place to identify issues quickly and take corrective action in a collaborative way.

From a management perspective, it is also clear that many decisions about the design of the capability cannot be taken by one manager alone. There needs to be the cross-functional, and sometimes cross-organisation, governance in place to ensure that the overall process is acceptable to all those parties involved.

Making this sort of change happen is difficult and it takes time and a reasonable amount of resource to do it well. It then takes a continuous resource level to sustain it successfully. You need to ensure that your organisation understands this and puts the right level of resource behind the change activity if it is to be successful.

What this translates into when it comes to designing artwork projects includes:

- Establishing a cross-functional (and sometimes cross-organisation) governance or steering team.
- Establishing a cross-functional project team.
- Identifying and involving the end-to-end process owner in the project.
- Establishing some clear design principles that can be used to guide the myriad of detailed design decisions that will need to be made.
- Performing process and capability design involving the cross-functional project team.
- Identifying other cross-functional representatives from across the organisation(s) to review and give input to proposed designs prior to them being finalised.
- Having a clear mechanism within the project to make design decisions that takes account of views from all the relevant parties and

provides appropriate escalation to the steering team if agreement cannot be reached.

- Communicating plans and proposals to the wider organisation(s) that will be impacted by the change as the project is proceeding.
- Actively identifying and managing issues that stakeholder groups may have with the design.
- Ensuring that education, training and competency assessment is thorough and robust.
- Ensuring that support processes and resources are also implemented to ensure that people executing the process remotely and infrequently will get the support they need when they need it.
- Performing a review of the project once completed to ensure all learning is captured, acknowledged and acted upon in future projects.

If you are thinking that this all sounds like design by committee, then let us make it clear that this is not what we are suggesting here at all. There is clearly a level of give and take required to ensure that everyone at least understands and buys into the resulting design. However, the design principles that you develop should guide decisions about the design and it is up to the project manager, and ultimately the steering team, to ensure this happens. In order to ensure this works effectively, it is also important to align the middle management of the organisation behind the concept and ensure they back decisions, even if they are not exactly what their staff might want.

The basic structure of the project is likely to be relatively familiar to you and we outline the usual phases in Figure 11.1.

The concept development phase has a number of key objectives:

- Get buy-in from key senior management that there is an issue that needs to be investigated.

- Form an initial steering team to govern the project.
- Design the situation and strategy phase of the project.
- Identify the resources required for the situation and strategy phase.
- Gain formal approval for the situation and strategy phase.

During this initiation phase of the project there is always a significant dilemma to manage. On the one hand, there is no project approved at this point, so there is no significant budget available and there are almost certainly only a very small number of part-time resources available to work on the project. On the other hand, management will always want a fully developed case for action and project design before moving forward. This latter desire cannot realistically be met until the end of the next phase of the project and usually requires a relatively significant amount of effort.

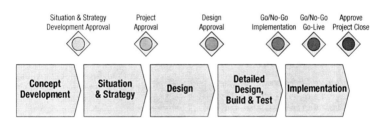

Figure 11.1 Typical artwork project phases

The situation and strategy phase is the time to develop the compelling reason for action, thoroughly understand the as-is situation and define a to-be at a high level. This needs to be done with the involvement and buy-in of the main impacted functions in the organisation, of which the steering team members are a key element. Having got the as-is and to-be, the gap between the two will be clear and the project can be designed, resources identified and cost estimates developed. If there are key suppliers to be selected, then this process should begin here. It may be advantageous or even necessary to complete the selection process in this phase for some or all of the key suppliers. This phase culminates in the approval of the resources and funds for the rest of the project and is therefore the critical project approval point.

The design phase of the project is concerned with collaboratively developing the to-be in more detail and also designing in detail all aspects of the remaining phases of the project. At the end of this phase, there should be a clearly developed to-be design, at a level of detail that all parties can understand and has been tested for robustness. As an example, this design would typically include: process flow maps; roles and responsibility tables; information technology user requirements and initial functional requirements. If there are key suppliers that need to be selected, then they will probably have been selected by the end of this phase and involved in the design. You will note that this design does not include writing standard operating procedures, education and training materials, or developing the detailed design of information technology systems. This phase would normally culminate in a formal approval of the to-be design by the steering team and any other key senior stakeholders who may be appropriate.

As the name suggests, the detailed design, build and test phase is the section of the project where the detailed design of the solution is completed, capabilities are built and tested. This will include items such as writing and approving standard operating procedures and education and training materials. Information technology systems and tools will also be designed, built and tested. It must be remembered that in a GxP and validated environment, the development and execution of the system testing is a significant endeavour in its own right, requiring significant quantities of resource and time to achieve.

Some time towards the end of this phase, a formal decision would normally be made that the design and tools are sufficiently well developed that the implementation phase can begin. The reason for this formal decision point is that the beginning of the implementation phase normally commits all impacted stakeholders to a significant amount of effort in preparation for "go-live" and therefore should only be commenced if the project is truly ready.

In our experience, many projects are delayed, or even fail, because managers underestimate the resources and time required to complete this phase adequately. This is often caused by too much optimism or bowing to pressure to cut timescales and cost in the situation and strategy phase. Ensuring some experienced, sufficiently senior views are brought to bear on plans in these early stages can help avoid this situation occurring.

Implementation, or deployment as some organisations call it, takes the detailed design, tools and information technology systems and implements them in the organisation. It is during this phase that the new processes and systems will "go-live" and be proven for real in the business. Typical activities in this phase would include deploying tools and information technology systems, data migration, training of staff and performing process qualification. We would recommend that there is a formal "go/no-go" decision made immediately prior to the first "go-live" of any new capability and that this decision is made by the steering team. This will ensure that the project and organisation are ready.

Having successfully achieved the "go-live", the job of the project team is not over. The project team needs to support the initial implementation of the new capabilities until they are stable and handed over effectively to the people who will operate them and support them on an ongoing basis. This phase of the project would normally include the following activities:

- Managing the cut-over from old capabilities to new capabilities.

- Providing increased support to users during the initial period after "go-live".
- Ramping up new capabilities to design capacity.
- Actively identifying issues in the new capabilities and rectifying them.
- Completing process qualification.
- Decommissioning redundant capabilities.
- Closing out all project activities and reporting.

Key roles

We will now take some time to discuss some of the roles associated with this type of project that we have found are key to its success. Figure 11.2 identifies the key roles we will discuss.

In our experience, all change projects need sponsorship at a senior level in the organisation to be successful. Appointing a senior member of staff as sponsor for a project will go a long way to ensure this sponsorship occurs effectively. This person needs to be senior enough in the organisation to be able to provide significant influence over all of the key functions impacted by the project. They will support the project at very senior levels in the organisation and help guide the project though any political issues it may face.

We have touched on the role of the process owner before. This is a person who has the end-to-end responsibility to ensure that the process works effectively for the organisation. Therefore, this person needs to have the skills and capabilities to manage and develop a complex business process on an ongoing basis. The cross-functional and cross-organisation nature of artwork processes means that they also need to have the skills, respect and seniority to manage the process effectively across many parts of the organisation over which they

Key Project Roles	Description of Role
Sponsor	• Executive ownership & support for project • Stakeholder engagement at highest levels in company • Provides strategic business direction
Process Owner / Business Lead	• Business leadership of project • Owns the resulting capability • Project team's immediate customer
Senior Subject Matter Expert	• Provides knowledge of artwork capability and improvement methodology best practice • Involved in strategy development; as-is assessment; root cause analysis; business case development; project design; supplier selection and solution design
Functional Representatives and Change Agents	• Represents their function on the project and contributes to many aspects of the project • Champions the project in their own function • Plays an active role in readying their function for the change
Project Manager	• Manages effective delivery of the project
Change Management Lead	• Defines change management approach on project • Ensures appropriate engagement of stakeholders at all levels • Ensures user readiness for change
IT Lead	• Manages all aspects of IT solution design and delivery
Business Analyst	• Develops IT business and functional requirements • Involved in detailed system design
Implementation Lead	• Responsible for implementation approach design and execution
Stream Leads	• Manage individual areas of the project
Key Supplier Leads	• Represents their organisation on the project and contributes to appropriate aspects of the project • Champions the project in their organisation • Responsible for readying their organisation for the change

Figure 11.2 Key project roles

have no line management jurisdiction. To be successful, they need to have the skills to manage and develop the process collaboratively with all impacted stakeholders. In many organisations, this role is also referred to as the project business owner.

The senior subject matter expert is a role that brings an overall knowledge of the artwork capability to the organisation. They should have a thorough understanding of artwork capabilities of all levels of capability, in order to help guide the organisation on the most appropriate solutions for them. Without this sort of input, many organisations will miss significant opportunities, make many avoidable mistakes and take significantly longer to achieve sustainable results than they otherwise could. More often than not, this role is external as it needs to bring an external perspective and experience to bear.

Functional leads represent each of the business functions and organisations impacted by the project. They play an active part in the project and the development of the future capability design with other members of the project team. They also play a key role as what we will call change agents. As change agents, they need to consult with their functions during the design to ensure that the views and experience of their function is fairly reflected in the new capability design. They also need to be champions of the change whenever they get an opportunity, helping to ensure that their function supports the project. If this is done well, it will significantly improve the chances of the future capability being accepted by the organisation.

It almost goes without saying that this sort of change initiative needs strong and experienced project management to be successful. The project manager needs to be chosen for their project management skills, particularly in multi-functional/organisation environments and to complement the change management skills available in the rest of the project senior team.

We would recommend dividing a complex project into a number of discrete blocks of activity or streams and assigning a stream lead to manage each area. Figure 11.3 lists some examples of project streams on a complex project. The stream lead is effectively a project manager for their area of responsibility. This is often an excellent opportunity for future project managers or department managers to develop their knowledge and skills.

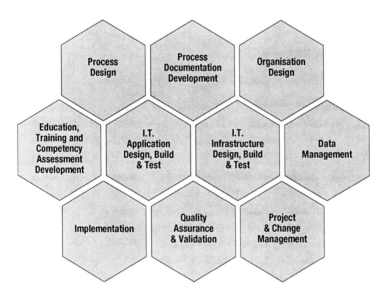

Figure 11.3 Typical project streams

Depending on the nature of the project, there may be significant external suppliers involved. They may be outsource suppliers who will have an ongoing involvement in delivering the overall artwork capability, or they may be suppliers who are just involved in the execution of the project. In all cases, lead representatives from these organisations should be invited to take active participation in the project and its management activities.

Help is out there

Getting the right external help is vital to the success of most major projects; artwork capability development activities are no exception.

We will discuss four groups of external help that are particularly pertinent to defining and implementing artwork capability improvements.

- Artwork capability subject matter experts.
- Outsource suppliers.
- Tools and technology suppliers.
- Transient project resource.

The nature of the internal development of an organisation's artwork capability often means that individuals involved in the capability are often technical specialists or relatively junior or both and have, more often than not, developed their artwork skills and knowledge within the company or with exposure to a very small number of other companies. Furthermore, there is no such thing as an artwork capability university, so opportunities to get an external perspective are limited. Therefore, as we have already suggested, an organisation can often benefit significantly from the help of external artwork capability subject matter expertise in the following areas:

- Understanding what good looks like in terms of a full scope artwork capability.
- Understanding how the different capability elements will bring benefit to their particular situation and business objectives and help design a suitable to-be.
- Helping to construct and sell the compelling case for change in the organisation.
- Helping architect a successful project.

- Independent assistance to identify and select appropriate artwork outsource and/or project suppliers.
- Helping in many areas of the design development.
- Coaching the organisation and project team through the difficult journey they face.

We have discussed outsource suppliers at great length in the outsourcing chapter and therefore will not go into any more detail on the topic here.

There are many opportunities for external suppliers to provide well developed tools and information technology solutions that bring with them the experience and best practice from work with other customers. We discussed this topic in the technology chapter and would refer you there for more details.

Projects, by their nature, are transient and most organisations do not have a permanent staff to meet all of their project team needs. Furthermore, just as managing operations requires specific skills and experience to do it successfully, project management and execution requires a different set of skills. Therefore, we would recommend organisations develop the appropriate skills and relationships to build project teams using a combination of a small number of internal operations and perhaps key project staff, together with a majority of external experienced project resource. We would suggest that internal resource is focussed on leading the change, ensuring that knowledge of the organisation is brought into the project. Project roles are challenging for everyone concerned, so try to ensure that you choose good people, not just those that are available. Projects provide an excellent opportunity to develop key talent in a short period of time.

Kotter's Steps for Implementing Successful Change	Some Thoughts on how this relates to Artwork Capability Improvement
Create a sense of urgency	• Clearly identify and communicate the compelling reasons for action • Develop the To-Be in a way that directly resolves key issues and links to delivering the overall business strategy • Phase capability improvement activity to focus on delivering shorter term, critical improvements first
Develop a guiding coalition	• Identify a senior business sponsor • Put in place a cross-functional governance group to steer the project and make key decisions for the business
Develop a vision for change	• Collaboratively develop a vision of the To-Be, involving the key impacted stakeholder groups
Communicate the vision	• Develop materials and messages to explain the To-Be vision in terms that can be understood by all impacted parties of the organisation • Use these materials and messages in many ways to communicate the vision to the broader population
Empower broad-based action	• Deliver the overall change though a number of individual initiatives or projects if possible • Ensure projects involve appropriately cross functional teams • Engage a broader group of stakeholders in key solution design activities
Generate short term wins	• Phase capability improvement activity to focus on delivering shorter term, critical improvements first • Consider picking off some "low hanging fruit" type benefits, delivering them and then publicising these wins • Publicise key progress milestones that occur on projects
Don't let up	• Ensure that the sponsor and steering team hold the projects and organisation accountable to deliver what has been promised and work through the inevitable issues that will occur
Make it stick in the organisation culture	• Ensure change initiatives have a focus on developing and supporting culture and behaviour that will help ensure the sustainability of the solutions they deliver • Maintain a cross functional governance body after the initial project(s) have finished with an accountability to maintain and develop the capability and culture that has been put in place

Figure 11.4 Elements of successful change

How change projects succeed

We will close this discussion on making the change happen by looking at some of the main reasons that change projects succeed. This essentially forms a high level checklist of all the things that you need to make sure are in place to ensure the success of your project.

For this purpose we will refer to research that Professor John Kotter performed at Harvard Business School. His research into the success and failure of major change initiatives in organisations led to the conclusion that change initiatives succeed when they do a number of specific things. Figure 11.4 looks at those key success activities and provides some of our recommendations on how they apply to improving an overall artwork capability.

We have found it useful to use this insight to help build a project scorecard that is reviewed by the steering team periodically to ensure that a project is covering all the critical areas to ensure success.

Close

Congratulations, you have reached the end of the book; we hope you have learned some new things and enjoyed the journey! By this time you should feel confident that you understand the main topics that we have covered. To summarise:

Packaging and artwork presents a significant compliance risk.

Packaging and artwork errors represent the single largest cause of packaging recalls in the pharmaceutical industry. For many companies, their current capabilities are struggling to meet today's requirements.

There are significant and immediate business challenges facing the pharmaceutical and healthcare industry.

Pharmaceutical and other healthcare companies are facing one of the most difficult periods in their history. Delivering many new product variants into as many different markets as quickly as possible is vital for their short- and long-term viability. At the same time, they need to enhance their reputation with regulators, governments, the public and other key stakeholders.

Packaging artwork management capabilities are critical to delivering this business strategy.

Essential to the strategy described above is the need to be able to develop, deliver and maintain a significantly increased number of packaging artwork designs for a rapidly expanding number of markets. For many companies, today's capabilities will not achieve this without significant improvement.

Delivering quality artwork is a complex endeavour involving many moving parts.

In the largest organisations, artwork capabilities involve thousands of people, working across many internal functions, in more than one hundred countries, involving tens, if not hundreds of external organisations. The capabilities require the skilful design and management of integrated business processes, organisations and facilities, which are enabled by a suite of sophisticated information technology systems. In smaller companies, whilst the scale is reduced, the fundamental challenges remain unchanged.

Establishing and delivering improvements in artwork capabilities is a significant, but achievable change management challenge.

Delivering change in this area requires the management of a complex interaction of business processes: people in many different functions, organisations and countries using many, often validated information technology tools. This requires careful and skilled project and change management skills to do it effectively if significant compliance risks are to be avoided.

Excellence is achievable.

We have transformed and managed the global artwork operations for one of the world's largest pharmaceutical companies, creating a world-class capability which has sustained industry leading performance over many years. We continue this journey by helping many of our clients today to deliver step change improvements in their artwork capabilities. This book shares much of the learning from this experience and we hope this will help you on your journey to deliver similarly excellent artwork capabilities.

As you would expect, we are constantly developing our thinking as we work with new clients and the business environment that they find themselves in changes. Follow us at any of the following links to stay in touch with these developments and find lots more valuable information:

www.be4ward.com

www.stephenmcindoe.com

www.andrewrlove.com

If you have any questions about artwork capabilities, or feel that you might need our assistance in understanding how to develop your organisation's artwork capabilities, please do not hesitate to contact us at enquiries@be4ward.com, or via one of the many social networks such as Facebook and Linked In that we are members of.

We have one final request. If you feel this book has been useful to you, please take a couple of minutes to go on to www.amazon.co.uk and recommend the book. This will help ensure that others get to benefit from this knowledge as well.

Thanks again for spending this time with us.

Stephen McIndoe   Andrew Love

Glossary of Terms

Term	Definition
Adherence	Carrying out the medicinal regime that has been agreed between patient and doctor accurately and on time.
Persistence	Maintaining adherence to a medicinal regime over an extended period of time.
Artwork	Text, graphics, symbols, coding and Braille information placed on packaging components.
Bill of Materials (BOM)	A list of the parts or components that are required to build a product.
Bleeds	Printing where the colour continues right up to the edge of the paper.

Capability	Elements such as business processes, people, information technology and third party service providers that come together to deliver a particular business result.
Colour separation	The act of decomposing an artwork into single-colour layers.
Electronic Printer Proof	A modified artwork file that can be used directly in the packaging component printing process.
Enterprise Resource Planning (ERP)	Integrated information systems which enable integrated management of a business or enterprise.
Good Document Practice (GDP)	Guidance which describes the way in which documents are created and maintained.
Good Manufacturing Practice (GMP)	Guidance that outlines the aspects of production and testing that can impact the quality of a product.
GxP	Used to describe any combination of GMP, GDP or any other good practice guidance which can impact product quality.
Information Technology (IT)	Technology used to manage and process information, including computer software and computer hardware.

Material Safety Data Sheet (MSDS)	A document containing information regarding the properties of a particular substance.
Native Files	The file format in which an application saves files by default.
Product Lifecycle Management (PLM)	The process of managing the entire lifecycle of a product from its conception, through design and manufacture, to service and disposal.
Pre-Press Activities	Changes to an electronic artwork file to turn it into images suitable for the production of printing press plates/films or for direct input to an electronic printing press.
Press Plates and films	Plates or films used in the printing presses to print the artwork image on packaging components.
Print Supplier	See *Printer*.
Printer	An organisation responsible for printing the physical packaging components. They often cut, emboss and glue the packaging materials as well.
Process Qualification	The measures taken to ensure that a process achieves the desired results consistently and correctly.
Serialisation	Uniquely identifying individual product packages.

Stakeholders	A person, group, organisation or system that affects or can be affected by an organisation's actions.
The Cloud	The delivery of computing as a service over the internet, rather than a product.
Trapping	Adjustment of an artwork to compensate for mis-registration between printing plates on a multi-colour printing press. Trapping involves creating overlaps (spreads) or underlaps (chokes) of objects during the pre-press process to eliminate mis-registration on the printing press.
Validated (IT solution)	Checking that the software meets its specifications and fulfils its intended purpose. This process is subject to comprehensive guidance from regulators such as the Food and Drug Administration (FDA) in the USA.
Visual Office	The use of visual tools to describe and manage activities in a business process or office.
Wet proof	A proof or sample component made on the printing press using the plates, ink and media (paper, board etc) which will be used for the final production components.
XML	A mark-up language that defines a set of rules for encoding documents in a format that is both human-readable and machine-readable.

About the Authors

Andrew Love

Andrew Love is a multi-award-winning packaging and artwork management strategist, leader and author who has a passion for improving patient safety. Andrew spent 10 years as head of global packaging design operations at GlaxoSmithKline, the world's second largest pharmaceutical company at the time.

During his time there he oversaw the transformation of the global artwork management activities into a world-class, award-winning capability.

Andrew is one of the founders of Be4ward which helps pharmaceutical companies and their supply base to improve patient safety and drive additional value from their product range. He now develops products for, and works with, a number of pharmaceutical and healthcare companies in achieving these aims.

Andrew is a professional engineer and MBA with over 20 years of experience working for the world's largest global companies.

Stephen McIndoe

Stephen McIndoe has a passion for improving patient safety and building optimal artwork management capabilities for clients, which have won many awards. Stephen worked with GlaxoSmithKline, the world's second largest pharmaceutical company at the time, as a management consultant over a period of many years, playing a leading role in the transformation of their global artwork management activities into a world-class, award-winning capability.

Stephen is one of the founders of Be4ward which helps pharmaceutical companies and their supply base to improve patient safety and drive additional value from their product range. He now develops products for, and works with, a number of pharmaceutical and healthcare companies in achieving these aims.

Stephen holds a first-class honours degree in engineering from a leading British university and is a professional engineer with over 20 years of experience working for the world's largest global life sciences companies as an employee and management consultant.

DATE DUE	RETURNED

CPSIA information can be obtained at www.ICGtesting.com
Printed in the USA
LVOW072012281112

309215LV00014B/646/P